A Topological Aperitif

Revised Edition

Stephen Huggett · David Jordan

A Topological Aperitif

Revised Edition

 Springer

Stephen Huggett
Plymouth, UK

David Jordan
Hull, UK

To Anne and Dinh

ISBN 978-1-84800-912-7   ISBN 978-1-84800-913-4 (eBook)
DOI 10.1007/978-1-84800-913-4

British Library Cataloguing in Publication Data
A catalogue record for this book is available from the British Library

Library of Congress Control Number: 2008937803

Mathematics Subject Classification (2000): 54-01, 57-01

Illustrations: Figures 4.12–4.14, 5.3–5.7, 5.21, 5.22, D.14, D.15 and D.18 are reproduced by the kind permission of the Polymorphic Construction Company Ltd. © 2000. All rights reserved.

Printed on acid-free paper

Springer Science+Business Media
springer.com

Foreword

Topology has been referred to as "rubber-sheet geometry". The name is apt, for the subject is concerned with properties of an object that would be preserved, no matter how much it is stretched, squashed, or distorted, so long as it is not in any way torn apart or glued together. One's first reaction might be that such an imprecise-sounding subject could hardly be part of serious mathematics, and would be unlikely to have applications beyond the amusement of simple parlour games. This reaction could hardly be further from the truth. Topology is one of the most important and broad-ranging disciplines of modern mathematics. It is a subject of great precision and of breadth of development. It has vastly many applications, some of great importance, ranging from particle physics to cosmology, and from hydrodynamics to algebra and number theory.

It is also a subject of great beauty and depth. To appreciate something of this, it is not necessary to delve into the more obscure aspects of mathematical formalism. For topology is, at least initially, a very visual subject. Some of its concepts apply to spaces of large numbers of dimensions, and therefore do not easily submit to reasoning that depends upon direct pictorial representation. But even in such cases, important insights can be obtained from the visual perusal of a simple geometrical configuration. Although much modern topology depends upon finely tuned abstract algebraic machinery of great mathematical sophistication, the underlying ideas are often very simple and can be appreciated by the examination of properties of elementary-looking drawings.

We find many examples of this kind of thing in this book. There are a great many diagrams, carefully chosen so as to bring out, in a directly visual way, most of the basic ingredients of topological theory. It provides a marvellous introduction to the subject, with many different tastes of ideas that can be appreciated by a reader without much in the way of mathematical sophistication. The reader who desires to follow up these fascinating ideas will stand in an excellent position to pursue the subject further, having mastered the basic techniques that are introduced here.

The authors of this work, Stephen Huggett and David Jordan, both have excellent credentials for explaining the beauties of this curiously austere but potentially enormously general form of geometry. Some 20 years ago, Stephen Huggett was a graduate student of mine, and he always had a particular flair for conveying the excitement that he himself felt for the magnificence of geometrical and topological forms of argument. David Jordan is known to me as the creator of some beautifully constructed and ingeniously precise geometrical shapes. Both authors are clearly well placed to do the job that they have set out here to do, and in this I believe that they have succeeded excellently.

Roger Penrose
Mathematical Institute, Oxford

Preface

Topology is geometry without distance or angle. The geometrical objects of study, not rigid but rather made of rubber or elastic, are especially stretchy.

We want to present mathematics that is mind-stretching and magic, of a style that is conceptual and geometric rather than formulaic. In doing so we hope to whet the reader's appetite for this way of thinking, which is at the same time very old and very modern. It started with classical Greek geometry and is still a key part of current mathematical research, which is especially lively in geometry and topology. Indeed, just as in classical Greece, our understanding of the physical universe depends upon this geometrical thinking.

The heart of the book is in the first five chapters: homeomorphisms, surfaces, and polyhedra. Although these ideas are broadly pitched at the level of a second year undergraduate, the authors expect a tenacious mind with much less background to grasp them. The arguments of Chapter 6, still geometrical in style, add strength to the earlier chapters. This is not a book of pure geometry, as is Euclid, but rests upon the fine structure of the real number system. These underpinnings are mostly extracted from the early chapters and collected in Appendix A. Appendix B gives a fleeting glimpse into knot theory, introducing the *Jones polynomial*. Further breadth is given in Appendix C, in which we sketch the curious and instructive early history of topology.

The main ideas are illuminated by a wide variety of geometrical examples that we hope will fascinate and intrigue. Although elementary, the mathematics in this book is sharp and subtle, and will not be properly grasped without serious attempts at the exercises, the essential challenge of which may be un-done by a premature glimpse of an illustrated solution. If you want to be a pianist you don't just read music and listen to it, you *play* it.

Several people have been extremely helpful to us in writing this book. We are very grateful to Colin Christopher, Neil Gordon, Charlotte Malcolmson, Dinh Phung, and Hannah Walker for all their work. Also, we are deeply indebted to John Moran for his patience with and dedication to the pictures.

For this revised edition we have made a number of corrections to the text and the figures, throughout the book, and we have written a short but com-pletely new section at the end of chapter four. The Klein bottle can be thought of as a sphere with a "Klein handle". We illustrate how, given a sphere with any number of ordinary handles and at least one Klein handle, all the ordinary handles can be converted into Klein handles. This is a part of the important "Classification Theorem" for surfaces.

Contents

Foreword   v

Preface vii

1. Homeomorphic Sets   1

2. Topological Properties 15

3. Equivalent Subsets 25

4. Surfaces and Spaces 51

5. Polyhedra 69

6. Winding Number 93

A. Continuity 105

B. Knots 113

C. History 121

D. Solutions 127

Bibliography 149

Index 151

1
Homeomorphic Sets

In this chapter we introduce the basic idea that lies behind all else in topology. We consider certain sets, mostly subsets of the plane or space, and seek understanding of what it means for two such sets to be *homeomorphic*, that is, the same in a certain topological sense.

The idea of being homeomorphic fits into a pattern with the ideas of congruence and similarity. Two sets are congruent if there is a correspondence between them such that the distance between any pair of points of one set is the same as the distance between the corresponding pair of points of the other set: in short, the correspondence is distance-preserving. Two sets are similar if there is an angle-preserving correspondence between them. In like manner, two sets will be homeomorphic if there is a correspondence between them, the requirement now being that of preserving only "closeness".

Until later, all sets under consideration are Euclidean sets, that is, subsets of the real line, the plane, space or, more generally, n-dimensional space $\mathbb{R}^n$ for some n. We define the idea of a closeness-preserving correspondence between such sets as follows.

Definition 1.1

We call a mapping f from S to T a *homeomorphism* if f is continuous and has a continuous inverse. We call S and T *homeomorphic* if there is a homeomorphism from S to T.

A recurring problem in topology is to decide whether or not a given pair of sets is homeomorphic. It is, in principle, straightforward to show that a pair

S. Huggett, D. Jordan, *A Topological Aperitif*,
DOI 10.1007/978-1-84800-913-4_1 © Springer-Verlag London Limited 2001, 2009

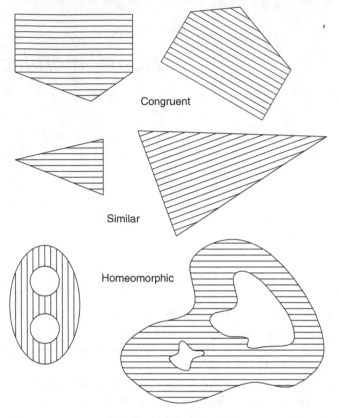

Figure 1.1

of sets is homeomorphic: an appropriate correspondence must be produced. In the next chapter, where our concern will be showing that sets are *not* homeomorphic, matters will be technically simpler but a little more subtle.

Example 1.1

Let S and T be the rectangles indicated in Figure 1.2. Start with S, stretch horizontally and then vertically, and we end up with T. So we define a mapping f from S to T by

$$f(x, y) = (2x, 3y),$$

with inverse

$$g(x, y) = (x/2, y/3).$$

Both f and its inverse g are continuous, so f is a homeomorphism, and S and T are homeomorphic. Of course there are many homeomorphisms

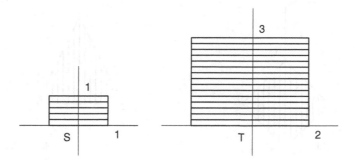

Figure 1.2 Example 1.1

from S to T, another being rotation of S by a right-angle followed by appropriate stretchings.

When defining a homeomorphism we prefer at this stage to give both a mapping and its inverse, so that the continuity of both can be seen. The homeomorphism of the previous example is neither area, length nor angle-preserving, so area, length and angle are not topological ideas.

Definition 1.2

A property is *topological* if it is preserved by homeomorphisms.

Thus, given a topological property and two homeomorphic sets, either both sets have the property, or neither has.

Our second example shows that a triangle and a square are homeomorphic, so that the number of corners of a set is not a topological property.

Example 1.2

Let S and T be the square and triangle indicated in Figure 1.3. The idea of the homeomorphism is to imagine the sets as made of thin vertical pieces and to push upwards, as shown. A simple formula gives this correspondence, namely

$$f(x, y) = (x, y + 1 - |x|)$$

with inverse

$$g(x, y) = (x, y - 1 + |x|).$$

As before, the continuity of f and g shows that S and T are homeomorphic.

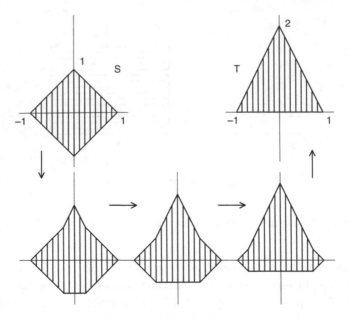

Figure 1.3 Example 1.2

Our next example is at this stage not immediately clear to the intuition. A line segment stretched to twice, or a hundred times, its length is homeomorphic to the original segment, but what if the segment is stretched infinitely?

Example 1.3

Let S be the open interval $]0, 1[$ of all points strictly between 0 and 1, and let T be the open interval $]1, \infty[$ of all numbers greater than 1. There is, surprisingly, a very simple formula for a homeomorphism f from S to T, namely, $x \rightarrow 1/x$ for $0 < x < 1$. This mapping is continuous on $]0, 1[$. The inverse of f is also $x \rightarrow 1/x$, but now for $x > 1$. This example shows that the property of being bounded is not a topological property, for we have two homeomorphic sets, of which one is bounded, but the other is not.

Figure 1.4 Example 1.3

Example 1.4

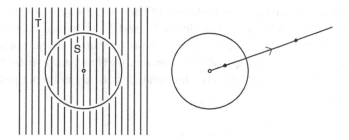

Figure 1.5 Example 1.4

This is a two-dimensional version of Example 1.3. By the unit circle S^1 we mean the set of points whose distance from the origin is 1, and by the open unit disc we mean the set of all points strictly inside S^1. Now let S be the open unit disc, with the origin removed, and let T consist of all points strictly outside S^1. To construct a homeomorphism f from S to T we map each point $\mathbf{x}$ of S radially outwards, using the idea of Example 1.3, so that the distance of $f(\mathbf{x})$ from the origin is $1/||\mathbf{x}||$, where $||\mathbf{x}||$ is the distance of $\mathbf{x}$ from the origin. Because the unit vector in the direction of $\mathbf{x}$ is $\mathbf{x}/||\mathbf{x}||$, we have

$$f(\mathbf{x}) = \frac{\mathbf{x}}{||\mathbf{x}||^2},$$

the inverse mapping, as before, having the same formula.

Example 1.5

The open unit disc is homeomorphic to the whole plane. As in the previous example we map points radially outwards, now sending a point whose distance from the origin is d to a point whose distance from the origin is $d/(1-d)$, giving the homeomorphism

$$\mathbf{x} \to \mathbf{x}/(1 - ||\mathbf{x}||),$$

with inverse

$$\mathbf{x} \to \mathbf{x}/(1 + ||\mathbf{x}||).$$

Note that the same formula provides a homeomorphism from the open unit ball, of all points of $\mathbb{R}^n$ whose distance from the origin is less than 1, to the whole of $\mathbb{R}^n$. In particular the open interval $]-1,1[$ is homeomorphic to the whole real line.

In the next two examples we make use of the *transitivity* property of being homeomorphic: if S is homeomorphic to T, and T is homeomorphic to V, then S is homeomorphic to V. We prefer to prove a little more, namely that being homeomorphic is an *equivalence relation*, so that sets can be put into classes—*equivalence classes*—sets in the same class being homeomorphic, and otherwise not.

Theorem 1.1

Being homeomorphic is an equivalence relation.

Proof

We must show that being homeomorphic is *reflexive*, that is, any set is homeomorphic to itself. This follows because the identity mapping from any set to itself is a homeomorphism. We must show that being homeomorphic is *symmetric*. This follows because, if S is homeomorphic to T, then there is some homeomorphism f from S to T, so that the inverse of f is a homeomorphism from T to S, and T is homeomorphic to S. Lastly, we show that being homeomorphic is *transitive*. Suppose that S is homeomorphic to T and that T is homeomorphic to V. Then there is a homeomorphism f from S to T, with inverse g say, and a homeomorphism h from T to V, with inverse k say. The mapping $s \to h(f(s))$ is continuous with continuous inverse $v \to g(k(v))$, so S is homeomorphic to V. This completes the proof.

Example 1.6

The open interval $]0,1[$ is homeomorphic to the whole real line. The previous example shows that $]-1,1[$ is homeomorphic to the real line. A homeomorphism from $]-1,1[$ to $]0,1[$ is given by $x \to (x+1)/2$, the inverse being $x \to 2x-1$. Consequently, appealing to Theorem 1.1, the interval $]0,1[$ is homeomorphic to the real line.

Example 1.7

We show that a square and a disc are homeomorphic. We take our square S to be $\{(x,y) : 0 < x,y < 1\}$, and D to be the open unit disc.

Example 1.6 assures us that there is some homeomorphism h from $]0,1[$ to the real line, so a homeomorphism from S to the plane is given by $(x,y) \to (h(x), h(y))$, the inverse being $(x,y) \to (h^{-1}(x), h^{-1}(y))$. But Example 1.5 tells us that the plane is homeomorphic to D. It follows, again using Theorem 1.1, that S and D are homeomorphic.

We leave the slightly harder case of a *closed* square and disc, that is including the edge points, as an exercise.

Example 1.8

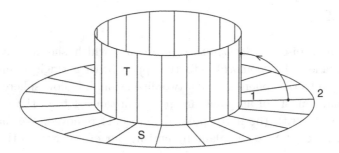

Figure 1.6 Example 1.8

We now have an example in three-dimensional space. Let S be the plane annulus $\{(x,y) : 1 \le x^2 + y^2 \le 4\}$ and let T be the cylinder $\{(x,y,z) : x^2 + y^2 = 1, 0 \le z \le 1\}$. Fold S upwards, shrinking its outer edge, and we end up with T. If we start with (x,y) in S, we want the corresponding point $f(x,y)$ in T to be in the same horizontal direction as (x,y), its distance from the z-axis to be 1, and its height equal to the distance of (x,y) from the unit circle, giving

$$f(x,y) = \left(\frac{x}{\sqrt{(x^2+y^2)}}, \frac{y}{\sqrt{(x^2+y^2)}}, \sqrt{(x^2+y^2)} - 1 \right).$$

If we start with (x,y,z) in T we want the corresponding point $g(x,y,z)$ in S to be in the same horizontal direction as (x,y,z) and to have distance $1+z$ from the origin, giving

$$g(x,y,z) = (x(1+z), y(1+z)).$$

Again, the continuity of f and g shows that S and T are homeomorphic.

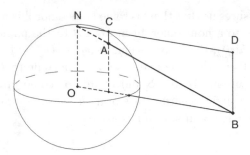

Figure 1.7 Example 1.9

Example 1.9

We now discuss *stereographic projection*, which shows that the whole plane is homeomorphic to the sphere with a single point removed. To us the sphere, S^2, is a two-dimensional, hollow sphere, not solid, and consists of all points in space a distance 1 from the origin. Let S be S^2 with the north pole $N = (0,0,1)$ removed. We map a point (x,y,z) of S to the plane by putting our eye at the north pole and "projecting" (x,y,z) on to the plane, giving us a point $f(x,y,z)$ of the form $(\lambda x, \lambda y)$. We now calculate λ. Let C and D be the points in the plane $z = 1$ above $A = (x,y,z)$ and $B = (\lambda x, \lambda y)$. Then

$$\frac{DB}{CA} = \frac{DN}{CN} = \lambda.$$

But DB is 1 and CA is $1 - z$, so we have $\lambda = 1/(1-z)$ and

$$f(x,y,z) = \left(\frac{x}{1-z}, \frac{y}{1-z} \right).$$

Note that f is continuous, because z is never 1. Continuing with our style of providing an inverse to justify our claim to have a homeomorphism, we find a formula for the inverse g of f. The formula f tells us that g must be of the form

$$g(x,y) = (x(1-z), y(1-z), z)$$

but we do not yet know what z is in terms of x and y. Now $g(x,y)$ lies on S^2, so we have

$$x^2(1-z)^2 + y^2(1-z)^2 + z^2 = 1,$$

and hence

$$x^2(1-z)^2 + y^2(1-z)^2 = (1+z)(1-z).$$

But z is not 1, so

$$x^2(1-z) + y^2(1-z) = 1 + z$$

which yields

$$z = \frac{x^2 + y^2 - 1}{x^2 + y^2 + 1},$$

giving

$$g(x,y) = \left(\frac{2x}{x^2 + y^2 + 1}, \frac{2y}{x^2 + y^2 + 1}, \frac{x^2 + y^2 - 1}{x^2 + y^2 + 1} \right).$$

We have shown that f and its inverse g are both continuous, so f is a homeomorphism, and S^2 with the north pole removed is homeomorphic to the plane.

Stereographic projection can very easily be adapted to other dimensions. For $n \geq 1$ we denote by S^n the n-dimensional sphere of all points in $\mathbb{R}^{n+1}$ a distance 1 from the origin. By the north pole, N, of S^n we mean $(0, \ldots, 0, 1)$. Stereographic projection from $S^1 \setminus \{N\}$ to the real line is

$$(x,y) \rightarrow \frac{x}{1-y}$$

Figure 1.8 Example 1.9

and from $S^3 \setminus \{N\}$ to space is

$$(x,y,z,w) \rightarrow \left(\frac{x}{1-w}, \frac{y}{1-w}, \frac{z}{1-w} \right).$$

For each $n \geq 1$, stereographic projection is a homeomorphism from $S^n \setminus \{N\} \rightarrow \mathbb{R}^n$.

The following little result is used in the next two examples and many times more.

Theorem 1.2

Homeomorphisms send subsets to homeomorphic sets.

Proof

Let f be a homeomorphism from S to T and suppose that X is a subset of S. We show that X is homeomorphic to its image $f(X)$. This is because the mapping from X to $f(X)$ given by $x \to f(x)$ is continuous, has the continuous inverse $y \to f^{-1}(y)$, and is consequently a homeomorphism. This completes the proof.

Example 1.10

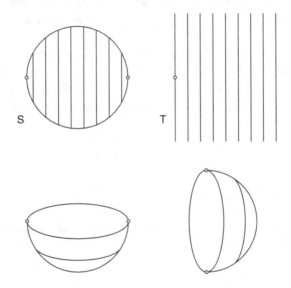

Figure 1.9   Example 1.10

We use stereographic projection to show that the plane sets $S = \{(x,y) : x^2 + y^2 \leq 1, x^2 \neq 1\}$ and $T = \{(x,y) : x^2 + y^2 > 0, x \geq 0\}$ are homeomorphic. The inverse, σ, of stereographic projection sends S to the southern hemisphere, including the equator, but excluding $(1,0,0)$ and $(-1,0,0)$. The image of T under σ is the eastern hemisphere, with north and south poles missing. Theorem 1.2 tells us that S is homeomorphic to $\sigma(S)$ and that T is homeomorphic to $\sigma(T)$. The rotation $(x,y,z) \to (x,-z,y)$ is a homeomorphism from $\sigma(S)$ to $\sigma(T)$. It now follows from Theorem 1.1 that S and T are homeomorphic. We observe that the problem is entirely two-dimensional, but the solution is three-dimensional.

Example 1.11

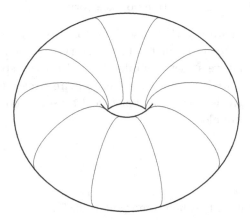

Figure 1.10 Example 1.11

As a final example of homeomorphism we discuss the torus. A torus is the surface of a ring doughnut, for which there is no very simple formula in three-dimensional space. To give a simple formula, and to emphasize the idea of a torus as a "product" of two circles, we define the torus in four-dimensional space, and use stereographic projection to show that our torus really is a torus.

Bearing in mind that the unit circle S^1 is $\{(x, y) : x^2 + y^2 = 1\}$ we define the torus T^2 as the "product" $\{(x, y, z, w) : x^2 + y^2 = 1, z^2 + w^2 = 1\}$ in $\mathbb{R}^4$. The torus therefore consists of all points of $\mathbb{R}^4$ of the form $(\cos u, \sin u, \cos v, \sin v)$. We go round a circle if we fix u and change v, or fix v and change u.

Every point of the torus is a distance $\sqrt{2}$ from the origin. So shrinking by a factor $\sqrt{2}$ gives us a subset of S^3, not including the north pole. Let T_S be the image under stereographic projection of our shrunk torus. Then, again using Theorems 1.1 and 1.2, the set T_S is homeomorphic to the torus T^2.

We calculate exactly what T_S is. Straightforward use of the formula from Example 1.9 shows that the image of

$$(\cos u, \sin u, \cos v, \sin v)/\sqrt{2}$$

under stereographic projection is the point $(1, 0, \cos v)/(\sqrt{2} - \sin v)$ rotated an angle u about the z-axis. A little more calculation shows that the set of all points of the form $(1, 0, \cos v)/(\sqrt{2} - \sin v)$ is the circle, lying in the x, z-plane, with centre $(\sqrt{2}, 0, 0)$ and radius 1. The set T_S is therefore the rotation about the z-axis of a circle in the x, z-plane, and really is what we think of as a torus.

In topology, when we say that a set is a torus, we mean that the set is homeomorphic to the torus T^2 first defined. Every torus is homeomorphic to every other torus, and we sometimes talk of "the" torus instead of "a" torus. Similarly, to us, a square is a circle. There are square circles and round circles: the word "circle" here is used in a topological sense, whereas "square" and "round" are outside topology.

EXERCISES

1.1. Prove that $[0, 1[$ is homeomorphic to $]3, 7]$.

1.2. Find an explicit formula for a homeomorphism from $]0, 1[$ to the real line.

1.3. Prove that the following two rectangles are homeomorphic:

$$\{(x, y) : 0 \le x \le 2 \text{ and } 0 \le y < 1\}$$

$$\{(x, y) : 0 \le x < 2 \text{ and } 0 \le y \le 1\}$$

1.4. Show that the plane sets $\{(x, y) : 0 < x^2 + y^2 < 1\}$ and $\{(x, y) : 1 < x^2 + y^2 < 4\}$ are homeomorphic. Let D be the closed disc $\{(x, y) : x^2 + y^2 \le 1\}$. Show that the plane with the origin removed is homeomorphic to the plane with the disc D removed.

1.5. Show that the cone

$$\{(x, y, z) : x^2 + y^2 = z^2, z \ge 0\}$$

is homeomorphic to the plane.

1.6. Show that the four subsets of space given below are homeomorphic.

The hyperboloid of one sheet given by $\{(x, y, z) : x^2 + y^2 = z^2 + 1\}$.

The cylinder $\{(x, y, z) : x^2 + y^2 = 1, |z| < 1\}$.

The cone $\{(x, y, z) : x^2 + y^2 = z^2, z > 0\}$.

The sphere with north and south poles removed.

1.7. Let C be the cube $\{(x, y, z) : 0 < x, y, z < 1\}$ and let B be the ball $\{(x, y, z) : x^2 + y^2 + z^2 < 1\}$. Show that B and C are homeomorphic.

1.8. Show that the three triangles below are homeomorphic.

$$\{(x,y) : 0 < y \le 1 - |x|\}$$

$$\{(x,y) : |x| - 1 < y \le 0\}$$

$$\{(x,y) : 0 \le y < 1 - |x|\}$$

1.9. Show that the closed square $\{(x,y) : |x|, |y| \le 1\}$ is homeomorphic to the closed unit disc $\{(x,y) : x^2 + y^2 \le 1\}$.

1.10. (This is more difficult.) By considering space as the union of spheres with centre the origin and using stereographic projection, or otherwise, show that spiked space $\mathbb{R}^3 \setminus \{(0,0,z) : z \ge 0\}$ is homeomorphic to the whole of $\mathbb{R}^3$.

2

Topological Properties

This chapter, in contrast to the last, shows that sets are *not* homeomorphic. To show that sets S and T are *not* homeomorphic we construct a suitable topological property P such that S has property P but T does not. From the definition of a topological property, if S were homeomorphic to T, then T would also have property P. So S and T are not homeomorphic. Thus every topological property is a tool for proving sets to be non-homeomorphic.

We give a collection of elementary topological properties that will be useful in the next chapter as well.

The topological properties we consider here are based on the intuitively clear idea of a *path*, which mathematicians tend to think of, not as a static object, but as a moving point. We denote by $[a, b]$ the closed interval of all x such that $a \leq x \leq b$.

Definition 2.1

A *path* α *in* S is a continuous mapping from the closed interval $[a, b]$ to S for some a, b where $a < b$. If $\alpha(a) = p$ and $\alpha(b) = q$, we say that α *joins p to q*.

Definition 2.2

A subset S of $\mathbb{R}^n$ is *path-connected* if every pair p, q of points in S can be joined by a path in S.

S. Huggett, D. Jordan, *A Topological Aperitif*,
DOI 10.1007/978-1-84800-913-4_2 © Springer-Verlag London Limited 2001, 2009

Example 2.1

The plane is path-connected. For let p, q be any points in the plane. The straight path given by

$$\alpha(u) = p + u(q - p), \quad 0 \le u \le 1,$$

joins p to q. Hence the plane is path-connected. There is nothing particularly two-dimensional about the argument, so we have also shown that $\mathbb{R}^n$ is path-connected for all $n \ge 1$.

Even with the sphere it would be getting complicated to write down explicit formulae for paths. Life is made easy for us by the following result.

Theorem 2.1

The continuous image of a path-connected set is path-connected.

Proof

Suppose that S is path-connected and that T is the image of S under the continuous mapping f. Take any points p, q in T. We must construct a path in T joining p to q. Now there are points x, y in S such that $f(x) = p$ and $f(y) = q$. Because S is path-connected, there is a path $\alpha : [a, b] \to S$ joining x and y. Define $\beta(t) = f(\alpha(t))$ for t in $[a, b]$. Then β is a continuous mapping from $[a, b]$ to T, and so is a path in T. Also $\beta(a) = p$ and $\beta(b) = q$, so β joins p to q. This completes the proof.

Example 2.2

The circle and the sphere are path-connected. The circle is the image of the real line under the continuous mapping $u \to (\cos u, \sin u)$, and the sphere is the continuous image of the plane under the continuous mapping $(u, v) \to (\sin u \cos v, \sin u \sin v, \cos u)$.

Example 2.3

The torus is path-connected, being the image of the plane under the continuous mapping

$$(u, v) \to (\cos u, \sin u, \cos v, \sin v).$$

Example 2.4

The punctured plane—the plane with the origin removed—is path-connected because it is the image of the whole plane under the continuous mapping $(x, y) \rightarrow e^x(\cos y, \sin y)$.

Theorem 2.2

Path-connectedness is a topological property.

Proof

Suppose that S is path-connected and that f is a homeomorphism from S to T. Then T is the image of S under the continuous mapping f so the path-connectedness of T follows from Theorem 2.1. This completes the proof.

Example 2.5

Let S be the real line with the origin removed. Any path in the real line from 1 to -1 must pass through the origin. So S is *not* path-connected. We now have our first proof that sets are not homeomorphic because S cannot be homeomorphic to the real line or any of the sets we showed to be path-connected.

Example 2.6

Figure 2.1   Example 2.6

Let T be the union of the half-axes $\{(0, y) : y > 0\}$ and $\{(x, 0) : x > 0\}$. Then T is *not* path-connected: any plane path joining points on these two axes must meet the line $y = x$. Alternatively, we could show that T is homeomorphic to the set S of Example 2.5.

We use path-connectedness to make another simple topological tool. The next step is to introduce language to express the idea that a non-path-connected set is formed of separate pieces, or *components*. Let S be a set. We say that points p, q in S are *together* if there is a path in S joining p to q. We wish to define the component containing the point p to consist of all points that are together with p, so we require the following result.

Theorem 2.3

Togetherness is an equivalence relation.

Proof

Let S be a set. Certainly, if p is in S, then p and p are together because mapping every point of $[0, 1]$ to p is a path in S joining p to p: thus togetherness is reflexive. Now suppose that p, q are together and that $\alpha : [a, b] \to S$ joins p to q. Go backwards along α: put $\omega(t) = \alpha(a + b - t)$. Then ω is a path in S joining q to p, so togetherness is symmetric. Finally, to show transitivity, let α join p to q as before and let β join q to r, where $\beta : [c, c + h] \to S$. Join p to r by going along α and then β. Define $\gamma : [a, b + h] \to S$ by

$$\gamma(t) = \alpha(t), \quad a \le t \le b$$

$$\gamma(t) = \beta(c + t - b), \quad b \le t \le b + h.$$

Then γ is a path in S joining p to r. This completes the proof.

Example 2.7

Figure 2.2   Example 2.7

Let T be the union of $\{(0, y) : y > 0\}$ and the whole real line. Then each point of T can be joined to the origin by a straight path in T. So each point is together with the origin, and, by transitivity, any two points in T are together. Thus T *is* path-connected.

Definition 2.3

A *component* is an equivalence class of togetherness.

Thus each point p of a set belongs to just one component, which consists of all those points that are together with p. To prove that sets with different numbers of components are not homeomorphic we need the following theorem.

Theorem 2.4

Homeomorphic sets have the same number of components.

Proof

Let f be a homeomorphism from S to T and suppose that α joins p to q in S. Then, as in Theorem 1, the path $t \to f(\alpha(t))$ joins $f(p)$ to $f(q)$ in T. Hence points that are together in S are sent to points that are together in T. Also points that are not together in S are sent to points that are not together in T, as otherwise the inverse of f would send points together in T to points not together in S. So the image of a component of S is a component of T, and the images of different components of S are different components of T. This completes the proof.

The previous result is not quite enough to deal with the following problems.

Example 2.8

Let S be the real line with the origin removed, and let T consist of those real numbers x such that $x < 0$ or $x = 1$. Then both S and T have two components. But each component of S is homeomorphic to $]0, 1[$, whereas T has one component homeomorphic to $]0, 1[$ and the other consisting of the single point 1. Now $]0, 1[$ and $\{1\}$ are not homeomorphic, for the superficial reason that they do not have the same number of points. Hence there is no way of pairing the components of S with those of T so that paired components are homeomorphic. The following theorem then tells us that S and T are not homeomorphic.

Theorem 2.5

The components of homeomorphic sets are homeomorphic in pairs.

Proof

Let f be a homeomorphism from S to T. Theorem 2.4 was proved by showing that a component of S can be paired with its image under f. But a set and its image under a homeomorphism are homeomorphic, so paired components are homeomorphic. This completes the proof.

Theorem 2.5 reduces the problem of deciding whether two sets are homeomorphic to the case where the sets are both path-connected.

Valuable information about a path-connected set is found by counting the number of pieces remaining when the set is "cut" by the removal of one point.

Definition 2.4

Let S be a path-connected set. We call a point p of S an *n-point* of S if removing p from S cuts S into n pieces, that is, $S\backslash\{p\}$ has n components. An n-point is also called a cut-point of type n, and a 1-point is called a not-cut-point.

Example 2.9

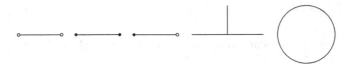

Figure 2.3   Example 2.9

Each point of the open interval $]0, 1[$ is a 2-point. The end points of $[0, 1]$ are 1-points, all other points being 2-points. The half-open interval $[0, 1[$ has one 1-point, all other points being 2-points. The set T of Example 2.7 has one 3-point, all other points being 2-points. The circle consists of not-cut-points.

Example 2.10

The set indicated in Figure 2.4 is path-connected, has infinitely many not-cut-points, but just one n-point for each $n \geq 2$.

For the calculation of Example 2.9 to provide proof that no two of the five sets are homeomorphic, we appeal to the next theorem.

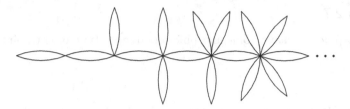

Figure 2.4   Example 2.10

Theorem 2.6

Homeomorphic sets have the same number of cut-points of each type.

Proof

Let f be a homeomorphism from S to T. We show that f sends each n-point of S to an n-point of T, so that, for each n, f gives a correspondence between the n-points of S and the n-points of T. Let p be an n-point of S. Then $S\backslash\{p\}$ has n components. But $S\backslash\{p\}$ is homeomorphic to its image $T\backslash\{f(p)\}$ under f. Consequently $S\backslash\{p\}$ and $T\backslash\{f(p)\}$ have the same number of components. So $T\backslash\{f(p)\}$ has n components, and $f(p)$ is an n-point. This completes the proof.

Example 2.11

Let S and T be the sets shown in Figure 2.5, the end points of the "arms" being missing. Both S and T have infinitely many 2-points and infinitely many not-cut-points. But the 2-points of S and T form the arms, including the points joining them to the circle. Hence the set of 2-points of S is path-connected, whereas the set of 2-points of T is not.

The next theorem gives the justification for saying that S and T are therefore not homeomorphic.

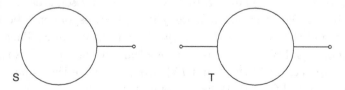

Figure 2.5   Example 2.11

Theorem 2.7

Homeomorphic sets have homeomorphic sets of each type of cut-point.

Proof

For any set X denote the set of n-points of X by X_n. Suppose that S and T are homeomorphic. We show that, for each n, S_n and T_n are homeomorphic. Let f be a homeomorphism from S to T. From the proof of Theorem 2.6 we know that f sends points in S_n to points in T_n and points not in S_n to points not in T_n. Hence the image of S_n under the homeomorphism f is T_n, and it follows that S_n and T_n are homeomorphic. This completes the proof.

Example 2.12

The circle and plane both consist of infinitely many not-cut-points. Removing any pair of points from the plane leaves a path-connected set, whereas removing any pair of points from the circle does not. To prove that the circle and the plane are not homeomorphic, we need to adapt our theory of cut-points to *cut-pairs*.

Definition 2.5

Let S be a path-connected set, and let p, q be distinct points of S. We call $\{p, q\}$ an *n-pair* of S if $S \backslash \{p, q\}$ has n components. An n-pair is also called a cut-pair of type n, and a 1-pair is called a not-cut-pair.

Theorem 2.8

Homeomorphic sets have the same number of cut-pairs of each type.

Proof

Let f be a homeomorphism from S to T. We show that f sends each n-pair of S to an n-pair of T. For each n, therefore, f gives a correspondence between the n-pairs of S and the n-pairs of T. Let $\{p, q\}$ be an n-pair of S. Then $S \backslash \{p, q\}$ has n components. But $S \backslash \{p, q\}$ is homeomorphic to its image $T \backslash \{f(p), f(q)\}$ under f. Consequently $S \backslash \{p, q\}$ and $T \backslash \{f(p), f(q)\}$ have the same number of components. So $T \backslash \{f(p), f(q)\}$ has n components, and $\{f(p), f(q)\}$ is an n-pair. This completes the proof.

EXERCISES

2.1. The plane set S is path-connected and is the union of three line segments, each segment being not only homeomorphic to $]0,1[$ but also straight. Find twelve examples of such a set S, no two of your examples being homeomorphic. Show that no two of your examples are homeomorphic.

2.2. The plane set S is path-connected and is the union of three line segments, each segment being not only homeomorphic to $[0,1]$ but also straight. Find eighteen examples of such a set S, no two of your examples being homeomorphic. Show that no two of your examples are homeomorphic.

2.3. The plane set S is path-connected and is the union of the axes and a circle (a round circle, not just a set homeomorphic to a circle). Find eight such sets S, no two being homeomorphic. Show that no two of the sets are homeomorphic.

2.4. The plane set S is path-connected and is the union of the vertical lines $\{(0,y) : 0 \le y \le 1\}$ and $\{(1,y) : 0 \le y \le 1\}$ and two horizontal closed line segments of length 1. Find eleven examples of such a set S, no two being homeomorphic. Show that no two of your sets are homeomorphic.

2.5. Let L_1 be the set

$$\{(x,0) : 0 \le x < 1\} \cup \{(0,y) : 0 \le y < 1\}$$

and let L_2 be congruent to L_1. The plane set T is $L_1 \cup L_2$. Sketch eight examples of such a set T, no two being homeomorphic. Show that no two of your examples are homeomorphic.

2.6. The sets S and T of Figure 2.5 have points where three lines emanate: in fact S has one whereas T has two. Give a precise definition of an *n-node*, a point where n lines emanate, and show that a homeomorphism sends an n-node to an n-node. The plane set X is path-connected and is the union of three circles (round circles, not just sets homeomorphic to a circle). Sketch eleven examples of such a set X, no two being homeomorphic. Show that no two of your examples are homeomorphic.

3
Equivalent Subsets

In this chapter we consider a development of the idea of homeomorphism, the various examples given making much use of the methods we now have for proving sets to be non-homeomorphic.

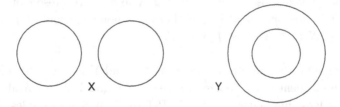

Figure 3.1

Consider the two plane sets X, Y of Figure 3.1. We know that the two sets are homeomorphic because each consists of two disjoint circles. We feel, however, that they are different in some topological sense. The way to see the difference is to consider the sets, not in isolation, but embedded in the whole plane. Think of X drawn in a plane made of our especially elastic topological rubber, and try to deform the plane so as to turn X into Y. We claim that it cannot be done: X and Y are essentially different as subsets of the plane.

In contrast to the plane, let us put the two sets in three-dimensional space, thought of as made of solid rubber. In this case we can deform one set into the other, dragging the rubber space with us.

S. Huggett, D. Jordan, *A Topological Aperitif*,
DOI 10.1007/978-1-84800-913-4_3 © Springer-Verlag London Limited 2001, 2009

This example helps us to see that what we are studying in this chapter is something more complicated than just a set. It is a set within a set. The two sets of Figure 3.1 are *equivalent* subsets in space, but non-equivalent subsets in the plane. To make a pair of circles non-equivalent to X in space we could link them, as in Figure 3.2, although this is actually a harder example: we postpone further discussion to Appendix B.

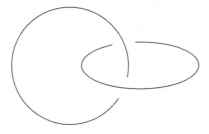

Figure 3.2

Our definition of the equivalence of subsets is given in terms of homeomorphism, so that, as with homeomorphism itself, we work with the idea of a correspondence rather than a deformation.

Definition 3.1

Suppose that S is a subset of $\mathbb{R}^n$, and that X and Y are subsets of S. We say that X and Y are *equivalent in S* if there is a homeomorphism f from S to itself sending X to Y, that is, $f(X) = Y$.

We emphasize that we are *not* defining "X and Y are equivalent" but "X and Y are equivalent in S". Whenever we say "X and Y are equivalent" there is always, at least implicitly, an "in S". Before discussing examples of subsets that are equivalent, and then subsets that are not, we give a simple consequence of the definition that perhaps helps to clarify the difference between the ideas of homeomorphic sets and equivalent subsets.

Theorem 3.1

Equivalent subsets are homeomorphic.

Proof

Suppose that f is a homeomorphism from S to itself that sends X to Y. We know that X is homeomorphic to its image $f(X)$ under f. But $f(X)$ is Y, so X is homeomorphic to Y. This completes the proof.

Example 3.1

In the plane, a circle is not equivalent to any interval. This follows from Theorem 3.1 because a circle, having no cut-points, is not homeomorphic to any interval.

Because of Theorem 3.1, from now on we ask whether X and Y are equivalent subsets only when X and Y are homeomorphic. We give a few examples of equivalent subsets before concentrating on non-equivalence.

Example 3.2

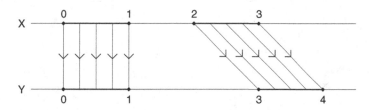

Figure 3.3 Example 3.2

Let X be $[0,1] \cup [2,3]$ and let Y be $[0,1] \cup [3,4]$. It is a simple matter to write down a homeomorphism from X to Y, namely

$$x \to \begin{cases} x, & 0 \le x \le 1 \\ x+1, & 2 \le x \le 3. \end{cases}$$

We have *not* yet shown that X and Y are equivalent in the real line: to

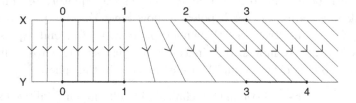

Figure 3.4 Example 3.2

do that we must define a suitable homeomorphism on the *whole* line, for example

$$x \to \begin{cases} x, & x \le 1 \\ 2x-1, & 1 \le x \le 2 \\ x+1, & x \ge 2. \end{cases}$$

Example 3.3

Let X be $[-1, 1[$ and let Y be $]-1, 1]$. Here X and Y are equivalent in the real line, because $x \to -x$ is a homeomorphism of the real line to itself sending X to Y. It is not enough here to think of deforming X to Y in an elastic real line: we must turn the whole line over.

Example 3.4

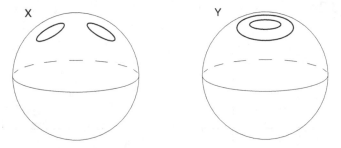

Figure 3.5   Example 3.4

Here X and Y both consist of two circles, and are the subsets of the sphere indicated in Figure 3.5. Somewhat surprisingly, especially in view of our remarks at the beginning of the chapter, X and Y *are* equivalent in the sphere. We will not give a formula but content ourselves with a description and picture of how to deform a sphere so that X becomes Y. In Figure 3.6 we move one of the circles of X to the south pole, expand it so that it becomes the equator, and finally shrink it towards the north pole: we now have Y. This process can be performed by stretching the whole sphere so that X ends up as Y. The whole sphere moves: the circles are not just sliding across a fixed sphere. Consequently, X and Y are equivalent in the sphere.

We allow ourselves a little more technical difficulty in our final two examples of pairs of subsets that *are* equivalent.

Example 3.5

We consider two sets, given in Figure 3.7, embedded in an open disc. As with the previous example, first impressions can be misleading: the subsets *are* equivalent. Let S be the open disc of all points whose

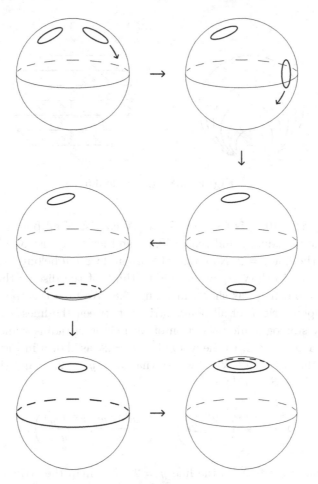

Figure 3.6 Example 3.4

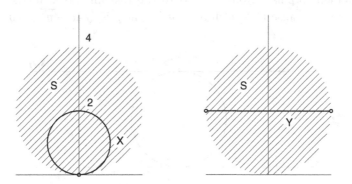

Figure 3.7 Example 3.5

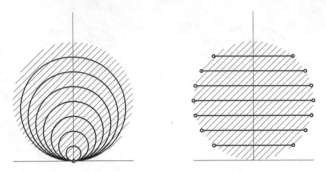

Figure 3.8 Example 3.5

distance from $(0, 2)$ is less than 2. Let X consist of all points of S on the circle of radius 1 and centre $(0, 1)$, and let Y consist of all points of S on the line $y = 2$. Note that the origin does *not* belong to X. The idea we use to show equivalence is to think of the disc as the union of circles and lines, as shown in Figure 3.8. We map the open disc S to the open strip T of all points strictly between the lines $y = 0$ and $y = 4$ by stereographic projection of each circle of radius k and centre $(0, k)$, for $0 < k < 2$, to the whole line $y = 2k$ as shown in Figure 3.9. We see that $x^2 + (y - k)^2 = k^2$, so that $x^2 + y^2 = 2ky$, and therefore define f from S to T by

$$f(x, y) = \left(x\frac{2k}{y}, 2k \right) = \left(x\frac{x^2 + y^2}{y^2}, \frac{x^2 + y^2}{y} \right).$$

Note that f sends X to the line $y = 2$. Now map the strip T to the disc S by shrinking horizontally, as follows. Let h be the inverse of the homeomorphism from $] - 1, 1[$ to the real line given in Example 1.5. Then, because the width of S at height y is $\sqrt{4y - y^2}$, a suitable

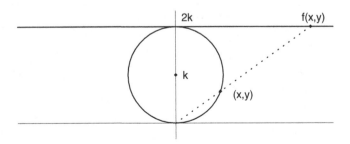

Figure 3.9 Example 3.5

shrinking is given by

$$(x, y) \rightarrow (h(x)\sqrt{(4y - y^2)}, y),$$

which sends the line $y = 2$ to Y. We have proved that X and Y are equivalent in S.

Example 3.6

Figure 3.10 gives five examples of a circle in a torus. Two of these show a circle going round once, but in different senses, two show a circle going round four times in one sense while going once round in the other sense, and the last, a trefoil knot, shows a circle going round twice in one sense and three times in the other. In Chapter 6 we will show that, keeping to the surface of the torus, none of the five circles pictured can be deformed into any other. We now show that every pair of these circles *is* equivalent in the torus. On the other hand the trefoil knot is *not* equivalent to any of the others as a subset of $\mathbb{R}^3$, because it is a proper knot. Some discussion of this harder result is given in Appendix B.

We must first articulate clearly what we mean by a circle going round the torus m times in one sense and n times in the other. Returning to our discussion of the torus T^2 in Example 1.11 we define $\tau(u, v)$ to be the point of T^2 that is the image under stereographic projection of

$$(\cos 2\pi u, \sin 2\pi u, \cos 2\pi v, \sin 2\pi v)/\sqrt{2}.$$

The points $\tau(u, v)$ and $\tau(u_1, v_1)$ are the same if and only if $u - u_1$ and $v - v_1$ are integers.

Suppose that $m, n \geq 0$ are relatively prime integers. Then the line segment in the plane joining $(0, 0)$ to (m, n) is mapped by τ to a (topological) circle on T, as the only points of the segment giving the same point of T are $(0, 0)$ and (m, n). We call this circle the (m, n)-circle. Thus the $v = 0$ circle is the $(1, 0)$-circle and the $u = 0$ circle is the $(0, 1)$-circle.

A continuous mapping g from the plane to itself gives a continuous mapping g^* from T to itself sending $\tau(u, v)$ to $\tau(g(u, v))$ provided that g^* is well defined, that is, that g sends all points in the lattice $\{(u, v) + (a, b) : a, b \in \mathbb{Z}\}$ to points in the lattice $\{g(u, v) + (k, l) : k, l \in \mathbb{Z}\}$ (see Appendix A). In particular $(u, v) \rightarrow (u, nu + v)$ gives a homeomorphism $T \rightarrow T$ that sends the $(1, 0)$-circle to the $(1, n)$-circle, so that the $(1, n)$-circle is equivalent in T to the $(1, 0)$-circle. To see what this

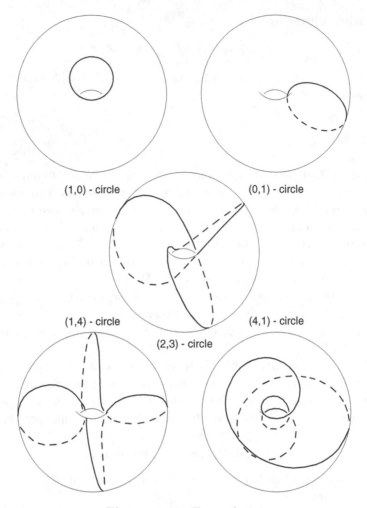

Figure 3.10   Example 3.6

homeomorphism of T to itself does, cut T round the $u = 0$ circle, wind one end round n times, and re-glue. Similarly $(u, v) \to (u+mv, v)$ shows that the $(0, 1)$-circle is equivalent in T to the $(m, 1)$-circle. Of course the $(1, 0)$-circle is equivalent in T to the $(0, 1)$-circle, a suitable homeomorphism from T to itself being given by the mapping $(u, v) \to (v, u)$ of the plane to itself.

To prove the general result that any two (m, n)-circles are equivalent in T we show that the (m, n)-circle is equivalent to the $(0, 1)$-circle. Define the mapping g from the plane to itself by $(u, v) \to (tu + mv, -su + nv)$, where s, t satisfy $ms + nt = 1$: we are assured that such s, t exist as m, n are relatively prime. As all four coefficients of u, v in the formula for g are integers, g does give a mapping $T \to T$, which is continuous and indeed a homeomorphism as the inverse of g is $(u, v) \to (nu - mv, su + tv)$.

Non-equivalence

We now change direction and concentrate on proving non-equivalence, starting with some simple, but very useful results.

Theorem 3.2

Equivalent subsets have equivalent complements.

Proof

Suppose that X and Y are equivalent subsets in S, and that f is a homeomorphism from S to itself that sends X to Y. We know that the complement $S \setminus X$ of X in S is homeomorphic to its image $f(S \setminus X)$. Because f is a bijection (a one-one and onto mapping) and the image of X is Y, it follows that $f(S \setminus X)$ is $S \setminus Y$. Consequently $S \setminus X$ and $S \setminus Y$ are equivalent in S. This completes the proof.

Combining Theorems 3.1 and 3.2 gives the following result.

Theorem 3.3

Equivalent subsets have homeomorphic complements.

Example 3.7

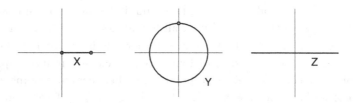

Figure 3.11   Example 3.7

We now give three plane sets, each homeomorphic to the open interval $]0,1[$, that are non-equivalent subsets in the plane. Let X be $]0,1[$, let Y be the unit circle with its north pole removed, and let Z be the whole real line. The complement of Z is not path-connected, whereas the complements of X and Y are. Also the complement of Y has a cut-point, namely the north pole, whereas the complement of X does not. Consequently no two of the three complements are homeomorphic, and Theorem 3.3 tells us that no two of X, Y and Z are equivalent in the plane.

A little later we will show that $]0,\infty[$, which is also homeomorphic to $]0,1[$, is not equivalent to any of X, Y, Z in the plane.

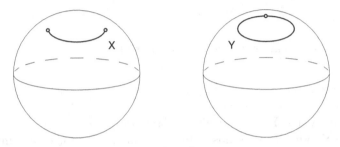

Figure 3.12   Example 3.7

An argument similar to the above shows that the sets X and Y shown in Figure 3.12 are non-equivalent in the sphere.

Example 3.8

Again we consider subsets of the sphere, but here each subset is homeomorphic to an open disc. Let X be the sphere with its north pole removed and let Y be the southern hemisphere excluding the equator. The complement of X consists of the north pole, whereas the complement of Y is a hemisphere, so the complements of X and Y are certainly not homeomorphic. From Theorem 3.3 we deduce that X and Y are non-equivalent in the sphere. Infinitely many non-equivalent subsets of the sphere can be found, all homeomorphic to an open disc. We give one more such subset. First, consider the shaded region, shown in Figure 3.13, that consists of two touching closed circular caps of the sphere, including their edge points and their common point. Our subset Z is the complement of this shaded region. As before, X and Z are non-equivalent in the sphere. The complement of Z is the shaded

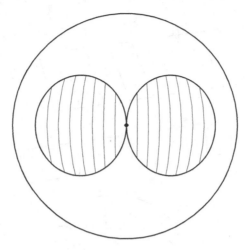

Figure 3.13   Example 3.8

region itself, which has a cut-point. Consequently Y and Z have non-homeomorphic complements, and it follows that Y and Z are non-equivalent in the sphere.

To help us in our next example we introduce an idealized representation of the torus, giving ourselves a simple way of drawing sets on the torus. First we take a rectangle, thought of as made of our usual stretchy topological rubber. Next we glue together two opposite edges to form a cylinder. Finally we glue together the ends of the cylinder to form a torus, as shown in Figure 3.14. This gluing is indicated in Figure 3.15 by drawing an arrow on each edge of the rectangle. The single arrows represent the formation of the cylinder and the double arrows represent the gluing of the cylinder ends to form the torus. Corresponding points on edges with the same arrow are to be thought of as the same. All four corners are the same.

Example 3.9

In this example we demonstrate non-equivalent ways of putting an open disc in a torus. As with the sphere, this can be done in infinitely many ways: we give four. Subsets X and Y are indicated in Figure 3.16. Note that Y does not go right round the torus: the points between p and q do *not* belong to Y. To see that Y really is an open disc, start with X, deform it to a rectangle, as shown in the two middle diagrams in Figure 3.16 and then expand the rectangle until it nearly meets itself. Expanding Y in the perpendicular direction yields the open disc Z

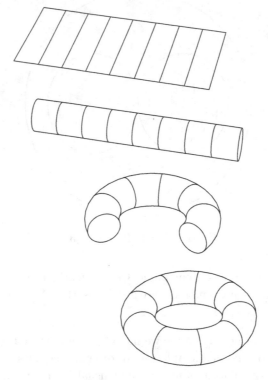

Figure 3.14

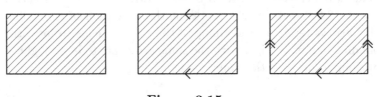

Figure 3.15

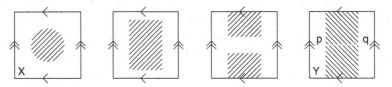

Figure 3.16   Example 3.9

shown in Figure 3.17, which also indicates a fourth subset W. We now

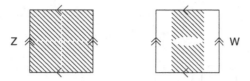

Figure 3.17 Example 3.9

study the complements of the four subsets in preparation for applying Theorem 3.3. The complement of Z in the torus is a figure 8, which has a single cut-point, whereas the complements of X, Y, W have no cut-points: note that the points between p and q are *not* cut-points of the complement of Y.

The complement of X has no cut-pairs, whereas the complement of Y has infinitely many cut-pairs, and the complement of W has just one cut-pair. Hence no two of the complements of X, Y, Z, W are homeomorphic and, by Theorem 3.3, no two of the sets X, Y, Z, W are equivalent in the torus.

Example 3.10

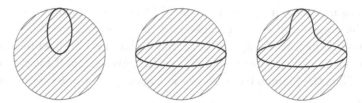

Figure 3.18 Example 3.10

In Example 3.4 we used the idea of deformation to help us see the equivalence of two subsets. In this example thinking of deformation can be misleading. Figure 3.18 shows three circles in a closed disc that can be deformed into each other. The three subsets, distinguished by the number of components of the complement, are non-equivalent: indeed there are infinitely many circles in the closed disc, no two equivalent, as we can find a circle whose complement has any desired number of components. Putting circles in a surface that *has* edge points makes little sense in equivalence questions. (Although the idea of an edge point seems clear, one has to be careful in making it precise, which we do in Chapter 6.)

Example 3.11

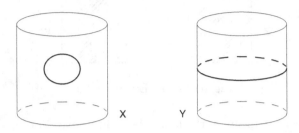

Figure 3.19 Example 3.11

We give two ways in which a circle can be a subset of a cylinder with its edge points excluded. Each set has a complement with two components, but one component of the complement of X is a disc, whereas both components of the complement of Y are cylinders, which we later prove cannot be homeomorphic to a disc. Consequently, by Theorem 3.3, the circles X and Y are non-equivalent in the cylinder.

Example 3.12

In this example we put circles in the *Möbius band,* shown in Figure 3.20, again excluding the edge points. In the same way that a cylinder can be represented as a rectangle with the ends glued together, as in the middle picture of Figure 3.15, a Möbius band is obtained by gluing together the ends of the rectangle, but first giving a half twist to one end. Thus the Möbius band, shown in Figure 3.20, can be represented as in Figure 3.21.

Three ways of putting a circle in the Möbius band are indicated in Figure 3.22: the circle goes round no times, once or twice, although we are not relying on the idea of "going round" for proof of non-equivalence. Note that Z really is a circle and not two circles. We can already prove that Y is not equivalent to either X or Z in the Möbius band, because the complement of Y is path-connected, whereas the complements of X and Z are not. As in Example 3.11, we will later show that the complement of X, one of whose components is a disc, cannot be homeomorphic to the complement of Z, whose components are a Möbius band (the central part) and a cylinder (the outer part). [If you have not already done this, take a Möbius band and cut it along the circle Z.] It follows that no two of X, Y and Z are equivalent in the Möbius band.

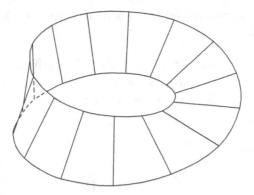

Figure 3.20   Example 3.12

Figure 3.21   Example 3.12

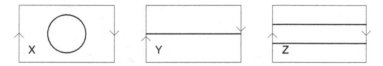

Figure 3.22   Example 3.12

There are some simple equivalence problems, one such being left over from Example 3.7, that we are not yet in a position to tackle. To solve these problems we introduce a new method, based on the idea of the *closure* of a set. To get the closure of a set we include the "edge" points as well. In the following, as with equivalence, we take a fixed set S and consider a subset X of S. The closure of X depends on S, but as the set S is fixed throughout a particular discussion, no confusion arises. In the definition below we use the word "neighbourhood", whose precise definition is given in Appendix A.

Definition 3.2

Let S be a Euclidean set and let X be a subset of S. The *closure* of X consists of those points s of S with the property that every neighbourhood of s meets

X, that is, every neighbourhood of s contains a point of X. We denote the closure of X by $\overline{X}$.

Of course, for any subset X of S, it follows that $\overline{X}$ contains X because, given any x in X, every neighbourhood of x meets X at x. Some examples of closure now follow. Let S be the plane and let X be the interval $]0,1[$ of the

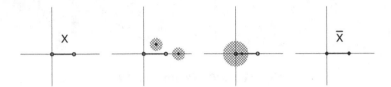

Figure 3.23

real line. If the point s neither belongs to, nor is an end-point of, the interval, then s has a neighbourhood *not* meeting X: see Figure 3.23. If s is an end-point of the interval, then *every* neighbourhood of s meets X: see Figure 3.23. Hence the closure of X in S is the interval $[0,1]$. Now let S be the open half-plane $\{(x,y) : x > 0\}$, and let X again be the interval $]0,1[$. Here the right-hand end-point of the interval is the only point in $\overline{X}$ but not in X: the origin is not

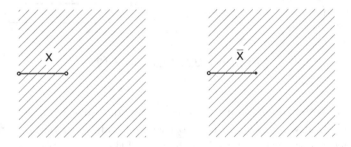

Figure 3.24

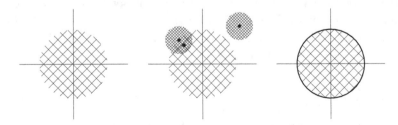

Figure 3.25

in S, and so not considered for $\overline{X}$. The closure of the open disc $\{x : ||x|| < 1\}$ in the plane is the closed disc $\{x : ||x|| \leq 1\}$.

It is perhaps appropriate, though not essential for the rest of this chapter, to make the following two general definitions at this stage.

Definition 3.3

A subset X of S is *closed* in S if the closure of X in S is X.

Definition 3.4

A subset X of S is *open* in S if every point in X has a neighbourhood contained in X.

Note that the half-open interval $[0, 1[$, as a subset of the real line, is neither open nor closed.

We now go ahead with our new equivalence result.

Theorem 3.4

Equivalent subsets have equivalent closures.

Proof

Let S be a Euclidean set, and let X and Y be equivalent subsets of S. Suppose that f is a homeomorphism from S to itself sending X to Y. We will show that f sends $\overline{X}$ to $\overline{Y}$. Take any s in $\overline{X}$. We first show that $f(s)$ belongs to $\overline{Y}$, so we consider any neighbourhood N of $f(s)$. Because f is continuous the pre-image M of N is a neighbourhood of s. But s is in $\overline{X}$, so there is some point x common to M and X. It follows that $f(x)$ belongs to N and Y. Hence $f(s)$ is in $\overline{Y}$, so that $f(\overline{X}) \subseteq \overline{Y}$. Similarly, $f^{-1}(\overline{Y}) \subseteq \overline{X}$, and we deduce that $f(\overline{X}) = \overline{Y}$. Thus $\overline{X}$ and $\overline{Y}$ are equivalent in S. This completes the proof.

Example 3.13

Let X, Y, Z be the plane sets given in Example 3.7, and let W be $]0, \infty[$, whose non-equivalence to X, Y, Z can now be shown. The closures of X, Y, Z and W have respectively $2, \infty, 0, 1$ not-cut-points, and so are non-homeomorphic. By Theorem 3.1, the closures of X, Y, Z and W are all non-equivalent subsets of the plane, and by the previous theorem, X, Y, Z, W are all non-equivalent plane sets.

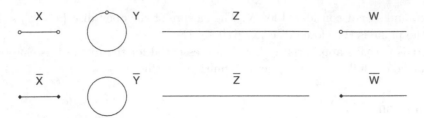

Figure 3.26   Example 3.13

Example 3.14

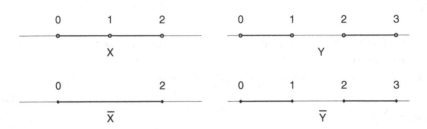

Figure 3.27   Example 3.14

Consider the sets $X =]0,1[\cup]1,2[$ and $Y =]0,1[\cup]2,3[$. Are the intervals of X stuck together, or is it possible to pull them apart by stretching the whole rubber sheet plane? Taking complements immediately shows that X and Y are non-equivalent in the real line. When we think of X and Y as plane sets, however, it is more helpful to use closure. Here $\overline{X}$ is path-connected, whereas $\overline{Y}$ is not, so $\overline{X}$ and $\overline{Y}$ are non-homeomorphic and therefore non-equivalent plane sets. Consequently X and Y are non-equivalent plane sets.

In the next example we combine the ideas of the previous two examples.

Example 3.15

We show that the seven subsets of the sphere indicated in Figure 3.28 are all non-equivalent.

The closures of A, B, C are not path-connected and have respectively 0,1,2 circular components. The subsets D, E, F, G have path-connected closures with respectively $0, 1, \infty, \infty$ 2-points, the closures of F, G having respectively $2, \infty$ not-cut-points. Hence the seven closures are non-homeomorphic and so non-equivalent in the sphere, from which it follows that the seven sets are non-equivalent in the sphere.

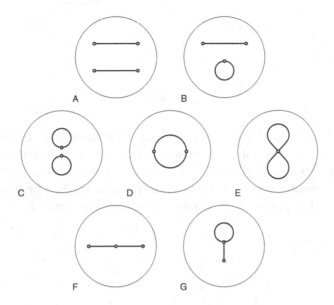

Figure 3.28 Example 3.15

Circles in a sphere

In Example 3.4 we showed that two apparently different ways of putting two circles in the sphere were equivalent. In this final section of the chapter we show how to put circles in the sphere in non-equivalent ways.

Figure 3.29 shows two ways of putting three circles in a sphere. These can be distinguished by an ad hoc argument: any two points of X can be joined by a path in the sphere, only the ends being in X, but this is not true for Y. The following discussion gives not just a method of showing non-equivalence but also a relatively simple method of counting how many non-equivalent ways four, five or more disjoint circles can be put in the sphere. In fact we prove only that there are *at least* two ways of putting three disjoint circles in the sphere. Similarly, for four or more circles, it will be proved only that there are *at least* the number of ways calculated.

The idea is to replace the problem of counting ways of putting a given number of circles in the sphere by a simpler, more algebraic, problem of counting so-called *trees*.

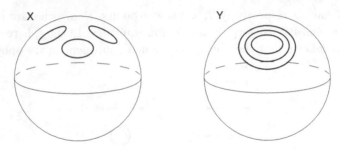

Figure 3.29

The first step is to define a *graph* (not in the usual sense) to consist of one set whose members we call *vertices,* and another set whose members we call *edges.* Each edge is related to, or *joins,* a pair of vertices (possibly the same). We shall represent vertices by points in the plane, and edges by lines between the corresponding points. Some graphs are indicated in Figure 3.30. A *string* is

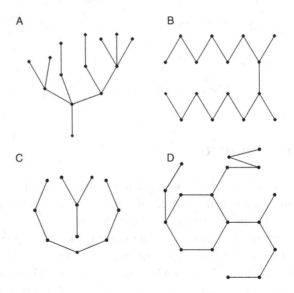

Figure 3.30

a finite sequence of alternating vertices and edges, each related to the next. We say that the ends of the string are *joined* by the string. A graph is *connected* if any two vertices can be joined by a string. A *circuit* is a string of distinct edges and vertices whose beginning and end are related. Thus D has a circuit, but A, B, C do not. A *tree* is a connected graph with no circuits. So A and B are trees, whereas C and D are not.

Analogously to the way that homeomorphic sets are regarded as the same, graphs with identical structure—*isomorphic* graphs—are regarded as the same. By an *isomorphism* between graphs we mean bijections from the vertices and edges of one graph to those of the other such that a pair and its image pair are either both related or both unrelated. Graphs are *isomorphic* if there is an isomorphism between them.

Any two trees having just two vertices are isomorphic: for short, there is only one tree with two vertices. This tree, the tree with three vertices, and the trees with four or five vertices are shown in Figure 3.31. It is straightforward to

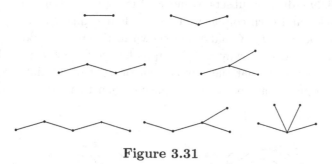

Figure 3.31

enumerate successively the trees having six or more vertices by systematically adding an extra vertex. There are six trees with six vertices, shown in Figure 3.32, and there are eleven trees with seven vertices.

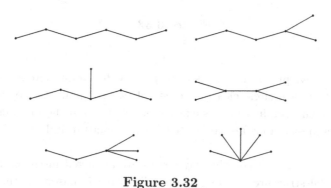

Figure 3.32

We next describe how a subset A of a set S gives rise to a graph, the *closeness* graph of A in S. We take as vertices one point from each component of A. Distinct vertices are joined by an edge if the corresponding components are *close*, an idea clarified below.

Definition 3.5

Let X and Y be subsets of S. We call X and Y *close* subsets of S if the closures of X and Y meet.

For example, the intervals $]0,1[$ and $]1,2[$ are close subsets of the real line because 1 is common to both closures, but $]0,1[$ and $]2,3[$ are not close.

Now we are in a position to construct our graph from a set of circles in the sphere, or the plane, by taking the closeness graph of the *complement* of the set of circles.

Figure 3.29 indicates subsets X and Y of the sphere, each subset consisting of three circles and each complement having four components. The closeness graphs corresponding to X and Y are shown in Figure 3.33. Both graphs are trees, and we now explain why such a graph, derived from circles in the sphere, is always a tree. Note that the complement of a set of disjoint circles in the sphere or the plane has one more component than there are circles: to avoid

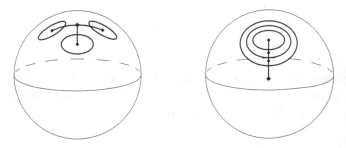

Figure 3.33

problems here we prefer our circles to be genuine flat round circles, so that the complement of each circle clearly consists of two components. So the number of vertices in the graph is always exactly one more than the number of edges. As such closeness graphs are connected, this happens if and only if the graph is a tree.

The connection between the way circles are arranged in the sphere or the plane and the structure of the corresponding tree is based on the following results.

Theorem 3.5

A continuous mapping from S to T sends close subsets of S to close subsets of T.

Proof

Let X and Y be close subsets of S, and let f be a continuous mapping from S to T. We must show that $f(X)$ and $f(Y)$ are close subsets of T. Because X and Y are close subsets of S there is some point s of S common to the closures of X and Y. Since s is in the closure of X, it follows that $f(s)$ is in the closure of $f(X)$, as was shown in the proof of Theorem 3.4. Similarly, $f(s)$ is in the closure of $f(Y)$, so that $f(s)$ is common to the closure of $f(X)$ and $f(Y)$. Hence $f(X)$ and $f(Y)$ are close subsets of T. This completes the proof.

Theorem 3.6

Equivalent subsets have isomorphic closeness graphs.

Proof

If X and Y are equivalent subsets of S, and f is a homeomorphism from S to itself sending X to Y, we know that f sends the components of X to those of Y. From Theorem 3.5 it follows that f sends close components to close components and f^{-1} sends close components to close components, so that the closeness relations of X and Y are isomorphic. This completes the proof.

Returning to the subsets X and Y of Figure 3.29, if these were equivalent, their complements would be equivalent and would have isomorphic closeness graphs. But the tree for X has a vertex related to three others, whereas the tree for Y does not, so the two trees are not isomorphic.

Take, for example, the case of putting five circles in the sphere. The six possible trees with six vertices have been given in Figure 3.32. Each such tree does give rise to a way of putting five circles in the sphere. Figure 3.34 shows how, by starting at a vertex and working away, we can construct circles yielding a prescribed tree. Consequently there are at least six ways of putting five circles in the sphere. We return to the problem we started with, of putting two circles in the plane. Why are there more ways of putting circles in the plane than in the sphere? Because, in the plane, one component of the complement is unbounded and, given a tree, the unbounded component can correspond to essentially different vertices of the tree.

By a *rooted* tree we mean a tree where one vertex, the root, has been chosen as special. Of course, by a *rooted isomorphism* we mean an isomorphism that sends a root to a root, and by *isomorphic rooted trees* we mean rooted trees with a rooted isomorphism between them. Figure 3.35 shows the two rooted trees with three vertices, the root being circled, and also the four rooted trees with four vertices. To construct the rooted tree given by a set of circles in the plane, first construct the tree as previously described, and then take as root the vertex corresponding to the unbounded component.

Although a continuous mapping does not always send a bounded set to a bounded set, a continuous mapping of the whole plane to itself *does* do so: see Appendix A. Now let f be a homeomorphism of the plane to itself sending X to Y, where X and Y are sets of circles. Then the only component of the complement of X that can be sent to the unbounded component of the complement of Y is the unbounded component. Hence the homeomorphism f gives rise to a rooted isomorphism of the rooted trees given by X and Y.

The two sets X and Y of Figure 3.1 give rise to the rooted trees A and B on the left of Figure 3.35. But A and B are not isomorphic rooted trees, for

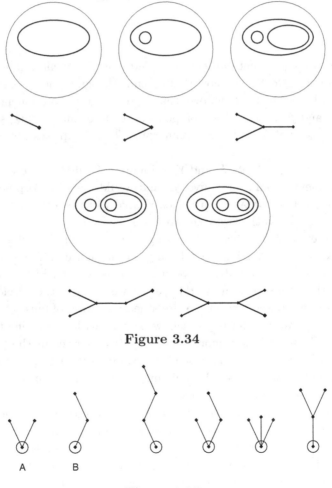

Figure 3.34

Figure 3.35

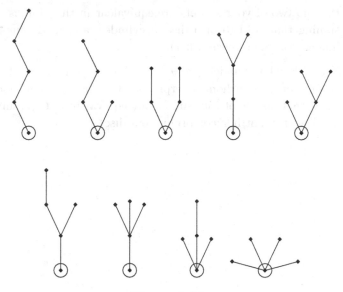

Figure 3.36

the root of A is related to two other vertices, whereas the root of B is not. Hence X and Y are non-equivalent subsets of the plane. Figure 3.36 shows the nine rooted trees with five vertices corresponding to the nine ways that four circles can be put in the plane. To construct circles in the plane yielding a given rooted tree, start with the root and work as before. Rooted trees with a given number of vertices can be counted by first enumerating the trees and then systematically rooting the trees in all possible ways.

EXERCISES

3.1. Let S be the union of the unit circle $\{(x,y) : x^2 + y^2 = 1\}$ with the line segment $\{(x,0) : -1 \leq x \leq 1\}$. Draw nine subsets of S, each homeomorphic to $]0,1[$, no two subsets being equivalent in S. Show that no two of your subsets are equivalent in S.

3.2. Find fourteen non-equivalent subsets of the sphere, each subset being homeomorphic to $]0,1[\cup]2,3[\cup]4,5[$ and having path-connected closure. Show that no two of your subsets are equivalent in the sphere: to complete the proof you will need a new argument.

3.3. Sketch eight non-equivalent subsets in the Möbius band (without edge points), each subset being homeomorphic to a figure 8. Show that no two of your subsets are equivalent in the Möbius band (assuming that no two of a disc, a cylinder, and a Möbius band are homeomorphic to each other).

3.4. Let C be the cylinder $\{(x, y, z) : x^2 + y^2 = 1, |z| < 1\}$. Find six subsets of C each homeomorphic to $]0, 1[$, no two equivalent in C. Show that no two of your subsets are equivalent in C (assuming that a cylinder is not homeomorphic to a disc).

4
Surfaces and Spaces

In this chapter we give a descriptive account of surfaces, of which we have already met the plane, the sphere and the torus. There are many other surfaces, shortly to be described. The essential idea is that near each of its points a surface is just like the plane.

Definition 4.1

A Euclidean set S is a *surface* if each of its points has a neighbourhood homeomorphic to an open disc.

A set consisting of two intersecting cylinders is *not* a surface: no point of intersection has a neighbourhood of the required form. For the disc, cylinder and Möbius band to be surfaces, we must leave off the edge points.

A closed cylinder, that is, a cylinder with its edge points but without its ends filled in to make a sphere, is not a surface but a surface with boundary. A Euclidean set S is a *surface with boundary* if every point of S has a neighbourhood homeomorphic either to an open disc or to the set $\{(x, y) : x^2 + y^2 < 1, x \geq 0\}$, shown in Figure 4.1. Closed discs, closed cylinders and closed Möbius bands are not surfaces, but *are* surfaces with boundary.

Our object in this chapter is to describe all those surfaces, like the sphere and the torus, that are bounded, closed and path-connected. To prove that we shall have listed all such surfaces—let us call them *good* surfaces—is beyond, but not far beyond, the scope of this book.

S. Huggett, D. Jordan, *A Topological Aperitif*,
DOI 10.1007/978-1-84800-913-4_4 © Springer-Verlag London Limited 2001, 2009

Figure 4.1

The sphere and the torus are the first of a sequence of good surfaces con-
structed by adding handles to, or equivalently making holes in, the sphere (see
Note 4.1). The torus is the sphere with one hole or handle: the two sets shown in
Figure 4.2 are homeomorphic. The double torus has two holes, or two handles.
More generally, the sphere with k holes in, or with k handles is the surface of
genus k. Each of these surfaces is *orientable*, a property that we describe only

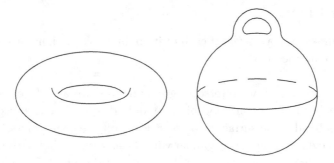

Figure 4.2

informally. To be orientable is to be two-sided. For example, the torus could
be painted red on the outside and green on the inside, and is two-sided, as is
the surface of genus k for each k. The Möbius band, though, is non-orientable,
or one-sided, for if you start painting a Möbius band red, you end up painting it all red. What gives spice to the study of surfaces is that there are good
surfaces other than spheres with handles. These new surfaces are considerably

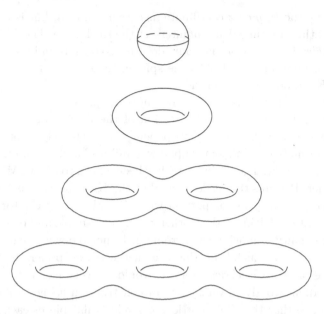

Figure 4.3

more difficult to describe, mainly because none of them occurs as a subset of three-dimensional space. Also, none of the new surfaces is orientable.

An alternative way of saying that a surface is non-orientable is if a triangular disc can be moved around the surface and returned to its original position with its sense changed. Note that we *do* need the filled-in triangle, which is why we called it a disc: if we simply drew the perimeter of a triangle (thus thinking of it as a topological circle) then even on a sphere we can reverse the sense.

The first of these new surfaces that we describe is the *Klein bottle*. We take a cylinder, bend the ends round, somewhat as in Figure 3.14, and join up the ends. However, there is a difference. We start with a tube thicker at one end than at the other, bend the thin end back, as shown in Figure 4.4, pull this thin end through the thicker part of the tube, enlarge the thin end and then glue it to the thick end. The perplexed topologist, having agreed that intersecting cylinders do not form a surface, wonders how the final set of Figure 4.4 can be a surface: it is not. To get rid of the intersection we first regard the set as lying in $\mathbb{R}^4$, with all points (x, y, z, w) in our set satisfying $w = 0$. We deform the thin part of the tube so that points at the intersection change from $(x, y, z, 0)$ to $(x, y, z, 1)$, and that, as we go along the thin part of the tube, the w-coordinates are first 0, steadily rise to 1 at the intersection, and finally come down to 0 again at the thick part of the tube. The resulting set in $\mathbb{R}^4$ *looks* the same as

before, because no x, y or z coordinate has been changed, but is now really a
Klein bottle: the thin tube does *not* intersect the thick tube. The Klein bottle is
non-orientable: if we start on the "outside" and go once round the Klein bottle
we end up on the "inside". Unlike the sphere, then, which has a well-defined
inside, the Klein bottle has no inside.

Now cut the Klein bottle into two congruent pieces by the plane passing
through the centre of each circular section of the tube. One of the resulting
pieces, shown in Figure 4.5, is a strip bent up at the edges like a piece of
gutter, given a half-turn and joined up: it is a Möbius band—again the apparent
intersection is illusory. The Klein bottle has been shown to be two Möbius bands
glued together. Perhaps the reader feels that the Klein bottle has been treated
unsatisfactorily vaguely in comparison with the sphere and the torus. We are
about to look at the Klein bottle in other ways, both simple and precise, but not
without the penalty of moving away from our hitherto down to earth approach
in which every set is Euclidean. However, if the reader prefers to stay in the
safe universe of these Euclidean sets, it would be very elementary, although
somewhat tedious, to give a version of our construction using a tube of square
cross section, so that the Klein bottle was made of flat pieces each given by a
straightforward formula.

As with the torus, the Klein bottle is a tube with the ends joined together,
and so can also be regarded as a rectangle with the edges glued together. To
make the Klein bottle, rather than the torus, one end of the tube is given a
half-turn rather than a whole turn before being glued and, analogously to the
Möbius band, can be represented as in Figure 4.6. We choose this opportunity
to make our jump away from the concrete and into the abstract by claiming
that the final picture of Figure 4.6 is not just a neat aid to thought but is a
clear indication of a precise mathematical object, not a Euclidean set but a
topological space. Whatever a topological space is, we would still like it to make
sense to ask whether two topological spaces are homeomorphic or not. But
being homeomorphic depends on continuity, which in turn depends on the idea
of neighbourhoods: a mapping f from a space X to a space Y is continuous at
the point $x \in X$ if the pre-image of N is a neighbourhood of x whenever N
is a neighbourhood of $f(x)$. Consequently we define a *topological space* to be
a set X with a specification of the neighbourhoods—subsets of X containing
x—of each point x of X. Certainly, therefore, Euclidean sets are topological
spaces. But wait a minute, says the reader, surely our neighbourhoods ought
to satisfy more axioms? Certainly, exhaustive treatises that begin with the
definition of a topological space, very often in terms of neighbourhoods, and
provide a list of more or less cryptic axioms, are manifold. Neighbourhoods
treated axiomatically are very fine if you intend to develop much abstract
theory, but we do not, and have no use for further axioms. Our purpose is

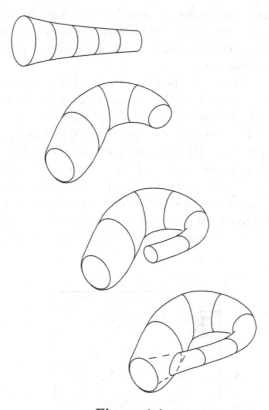

Figure 4.4

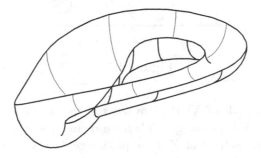

Figure 4.5

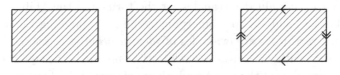

Figure 4.6

to make mathematically correct what at first may seem nebulous topological entities, for example, the glued rectangle of Figure 4.6. Nevertheless, every topological space we meet will satisfy all the usual axioms and no conflict with axiomatic topology will arise.

What exactly do we mean then, by our glued rectangle? To fix ideas, we consider the glued square X of Figure 4.7. What is X? There are points of different types in X. Whenever $0 < x, y < 1$, take $\{(x,y)\}$ as a point in X. Whenever $0 < x < 1$, we put the plane points $(x,0)$ and $(x,1)$ together to form the single point $\{(x,0),(x,1)\}$ of X. Similarly, whenever $0 < y < 1$, we take $\{(0,y),(1,1-y)\}$ as a point of X and, lastly, take $\{(0,0),(1,0),(0,1),(1,1)\}$ as a single point of X. We prefer to have $\{(x,y)\}$, rather than just (x,y) in X, so that every point of the original square *belongs* to a point of X, and the gluing, or identifying, of points of the original square can be regarded as an equivalence relation whose equivalence classes are the points of X. What

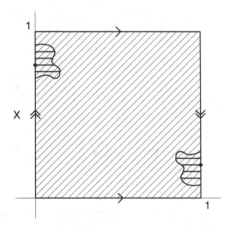

Figure 4.7

are the neighbourhoods of X? If s is a point of the original square, denote by $f(s)$ the point of X containing s. Then we define a subset N of X to be a neighbourhood of a point p of X if the pre-image $f^{-1}(N)$ is a neighbourhood of each s such that $f(s) = p$. A neighbourhood of one of the glued points of X has been indicated in Figure 4.7. We have made X into a topological space. Similarly, our earlier glued rectangles for the torus and the Möbius band are topological spaces and are homeomorphic to the corresponding Euclidean sets.

Our glued rectangles are examples of *identification spaces*. (See Note 4.2.) Let S be a space with an equivalence relation. Denote by $\hat{s}$ the equivalence class containing s and denote the set of equivalence classes by $\hat{S}$. The set $\hat{S}$ becomes an *identification space* by taking a subset N of $\hat{S}$ to be a neighbourhood of $\hat{s}$ if the union of all the classes in N is a neighbourhood of s whenever s is in $\hat{s}$. A further example of an identification space that we have essentially met before is the torus identified from the (whole) plane, the equivalence relation on the plane relating (u, v) with (u_1, v_1) if and only if $u - u_1$ and $v - v_1$ are integers: a detailed discussion is in Appendix A.

We now deem that all earlier definitions have been generalized to apply to topological spaces rather than just Euclidean sets. In particular, a *surface* is a topological space in which each point has a neighbourhood homeomorphic to an open disc. A *good* surface is a topological space homeomorphic to a closed, bounded, and path-connected Euclidean space. This adjustment to our definition of a good surface is not quite cost free: our terminology would be perverse without the fact, proved in Appendix A, that a Euclidean set that is a good surface under the new definition is indeed closed and bounded.

Our next non-orientable surface is called the real projective plane, whose peculiar name we hope to explain. Our initial view of this surface will be in terms of the cross cap, which is just a Möbius band whose boundary is the unit circle, no other points of the Möbius band lying in the plane of the circle. If you can't visualize this, don't worry, for you cannot construct a cross cap in three-dimensional space. Consider the Möbius band of Figure 4.8. Join each point p

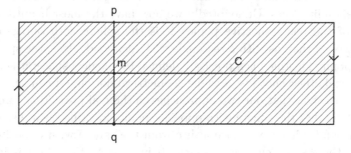

Figure 4.8

of the boundary circle to the opposite boundary point q by a straight line. The Möbius band consists of these lines, no two of which meet. The mid-points of the lines form a circle C and, as p moves once round the boundary circle, the corresponding mid-point m moves round C *twice*. Keeping these properties in mind, we construct our cross cap by joining each point $p = (\cos\theta, \sin\theta, 0, 0)$ of the unit circle to the point $m = (0, 0, \cos 2\theta, \sin 2\theta)$ by a line segment. The only points of the form $(x, y, 0, 0)$ in the cross cap are those on the unit circle. Consequently a surface, the real projective plane, is constructed by gluing the cross cap to the unit disc $\{(x, y, 0, 0) : x^2 + y^2 \leq 1\}$. The cross cap, being homeomorphic to the Möbius band, is non-orientable. Hence the real projective plane, which contains a cross cap, must be non-orientable.

To sum up this first approach to our new surface in one phrase, it is that the real projective plane is a Möbius band glued to a disc. In a sense, therefore, the real projective plane is simpler than the Klein bottle, which is two Möbius bands glued together.

For a different view of the real projective plane, we consider the meaning of the expression "real projective plane". Geometrically, *points* in the real projective plane can be thought of as *lines* in space that go through the origin, and a *line* of the real projective plane consists of all "points" lying in a plane through the origin. In this way *any* two lines of the real projective plane meet in a point, whereas in Euclidean geometry some lines—parallel lines—do not meet. Let p be a point of the real projective plane. Then N is a *neighbourhood* of p if N contains all points q whose angle from p is less than α, for some positive α. If we define the "distance" between two points p, q of the real projective plane to be the angle between the lines p, q then the definition of neighbourhood we gave for Euclidean sets still makes sense, so this set of points has been made into a topological space.

Next we slightly alter our way of thinking by replacing each line of space that goes through the origin by the pair of points where the line meets the unit sphere S^2. The "distance" between the two such pairs is the angle between the corresponding lines. Undoubtedly we now have the most beautiful way of thinking of the real projective plane—it is the sphere with opposite points identified. Note that, as with a rectangle with identified edges, the sphere with opposite points identified is a precisely defined topological space: it is a proper mathematical object, not just a vague geometrical notion. Of course, the problem now is to show that our two approaches to the real projective plane agree. We will show that the sphere with opposite points identified is a disc glued to a Möbius band. First we remove a disc from the sphere with opposite points identified, the disc we choose to remove being the Arctic, which is identified with the Antarctic, leaving the cylinder shown in Figure 4.9. Remember that opposite points are still identified. Because each point of the front half of this cylinder is identified with a point of the back half, we lose nothing by removing the back half. The only opposite points remaining are the left and right edges, and these are identified as indicated by arrows in Figure 4.9. But Figure 4.9 shows a strip with the ends identified to give a Möbius band. So the sphere with opposite points identified is the real projective plane.

Starting again with the sphere with opposite points identified and removing the open southern hemisphere, we see that the real projective plane can also be regarded as the northern hemisphere with opposite points of the equator identified. Now squash the hemisphere flat: the real projective plane is seen to be a disc with opposite points identified. Finally, turning the disc into a square

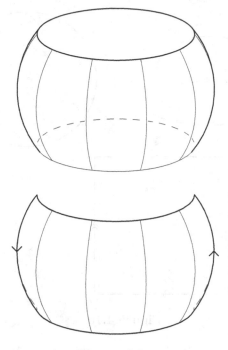

Figure 4.9

gives the representation of the real projective plane as a square with opposite points identified, the process being shown in Figure 4.10.

The reader may note that we now have square representations of the torus, the Klein bottle, and the real projective plane. As a slight digression, it is worth pointing out that the only other surface that can be so represented is the sphere, as in Figure 4.11.

Taking the view that, among the non-orientable surfaces, the real projective plane is important but difficult to understand, we give a third approach in terms of a more straightforward Euclidean set. We rather anticipate the next chapter—on polyhedra—by starting our construction with the octahedron, consisting of the eight equilateral triangles shown in Figure 4.12. If the reader wishes to be more specific the six vertices may be taken as the points lying on the axes and having distance 1 from the origin. Remove alternate triangles, leaving the four shaded triangles of Figure 4.12. Each coordinate plane meets the octahedron in a square, for example the x, y-plane meets the octahedron in the square with vertices $(1, 0, 0, 0), (-1, 0, 0, 0), (0, 1, 0, 0), (0, -1, 0, 0)$. Take these three squares, not just the edges but sets homeomorphic to a closed disc, *and* the four triangles. All we have to do now to get a surface is to dispose of the intersections using the same device as for the Klein bottle, namely by

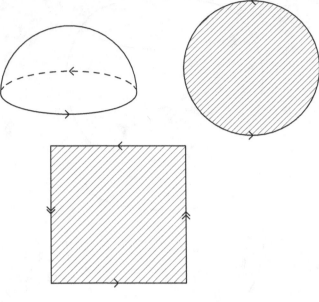

Figure 4.10

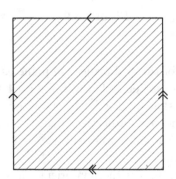

Figure 4.11

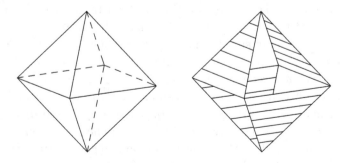

Figure 4.12

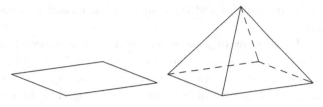

Figure 4.13

thinking of the set in four-dimensional space. We move the centre of two of the squares to give pyramids, as in Figure 4.13. The centre of the square in the x, z-plane is moved to $(0, 0, 0, 1)$, and the centre of the square in the y, z-plane is moved to $(0, 0, 0, -1)$. The square in the x, y-plane is not moved. As with the Klein bottle, our set *looks* the same as before because no point (x, y, z, w) has had x, y or z changed. But now no two of the squares meet except at the edges, for the interiors of the squares consist of points satisfying $w > 0$, $w < 0$ and $w = 0$ respectively. We must show that the new surface is indeed the real projective plane. We remove one square, labelled $AECF$ in Figure 4.14. The remaining two squares and four triangles are connected together as shown (with elongated triangles) in Figure 4.14, and form a Möbius band. Consequently our

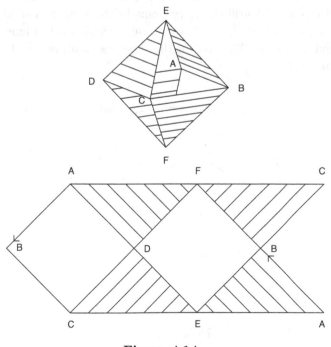

Figure 4.14

construction of four triangles and three squares has indeed produced the real
projective plane.

A disc can be thought of as a sphere with a disc removed, so the real
projective plane, which is a disc glued to a cross cap, can be thought of as
a sphere with a disc removed and replaced by a cross cap. We can construct
the Klein bottle, which is two Möbius bands glued together, by removing the
northern and southern hemisphere of a sphere, leaving just the equator, and
filling in both hemispheres with Möbius bands, or cross caps. Thus the real
projective plane is a sphere with one cross cap, and the Klein bottle is a sphere
with two cross caps. A non-orientable surface is constructed by taking a sphere,
removing one or more discs, and filling in with cross caps—the sphere with n
cross caps for some n. The real projective plane and the Klein bottle are the
cases $n = 1, 2$. It can be shown that every good surface belongs to one of our
two sequences of surfaces: every good orientable surface is either the sphere or
the sphere with a number of handles, and every good non-orientable surface is
a sphere with a number of cross caps.

We return for a further view of the Klein bottle, showing us a close rela-
tionship between the sequences of orientable and non-orientable surfaces. Take
a torus constructed symmetrically about the origin, for example the torus T_S
of Chapter 1, so that for every point (x, y, z) in the torus, there is an opposite
point $(-x, -y, -z)$ in the torus. Identify opposite points of the torus, and we
have a surface. To see what surface we have, we treat it as we did the sphere
with opposite points identified. Every point in the front half of the torus is
identified with a point at the back, so our surface, shown in Figure 4.15, can
be considered as a tube. The ends of the tube are identified as shown, so we
have a Klein bottle.

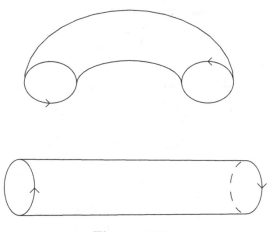

Figure 4.15

The sphere with any number of handles can be constructed symmetrically about the origin, in which position it is meaningful to identify opposite points. In fact, identifying opposite points on such a sphere with n handles, for $n \geq 0$, gives a sphere with $n + 1$ cross caps, though this general result is not as geometrically clear as the two cases we have considered. Thus the good surfaces can be thought of as either a sphere with n handles, for $n \geq 0$, or such a surface with opposite points identified, the sphere with $n + 1$ cross caps, for $n \geq 0$.

We do not here prove the Classification Theorem for surfaces, namely that each good surface is a sphere with handles or a sphere with cross caps, but we conclude this chapter with a geometric answer to the question: why don't we get a lot of new surfaces by having spheres with several handles *and* cross caps?

Our discussion starts with a new picture of the Klein bottle, shown in Figure 4.16, in which form we think of the surface as a sphere with what we call a Klein handle. So we can think of a pair of cross caps on a sphere as being a sphere with a Klein handle. The distinction between a Klein handle and a real handle is meaningful only if the handle is attached to an orientable surface. The handles are distinguished by the relative orientation of the ends of the handle—a cylinder—on the orientable surface: opposite for a real handle,

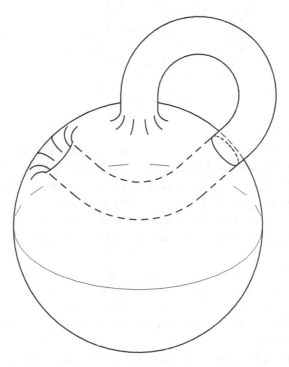

Figure 4.16

the same for a Klein handle. With this in mind the counter-intuitive result following is less surprising.

Figure 4.17 shows a sphere with a Klein handle and a real handle, and Figure 4.18 shows a sphere with two Klein handles.

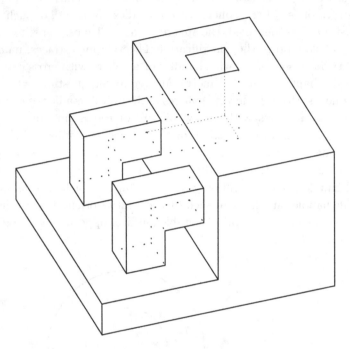

Figure 4.17

We show that the two surfaces are homeomorphic. The result of shrinking one of the Klein handles of Figure 4.18 is shown in Figure 4.19, this brief description being taken as an indication that the surfaces of Figures 4.18 and 4.19 are homeomorphic. Continuing in this way, our description always to be taken as indicative of a homeomorphism, we move an end of our shrunk Klein handle along and down the side of the other Klein handle, as shown in Figure 4.20.

Moving the end of the shrunk Klein handle further along the other Klein handle gives the surfaces shown in Figures 4.21 and 4.22.

Finally, moving the end of the shrunk Klein handle completely off the end of the other Klein handle, shown in Figure 4.23, the shrunk Klein handle has become a real handle.

Thus, given a sphere with any number of real handles and at least one Klein handle, all the real handles can be converted into Klein handles.

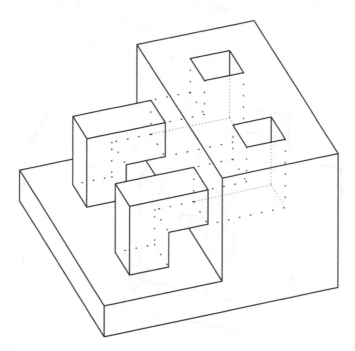

Figure 4.18

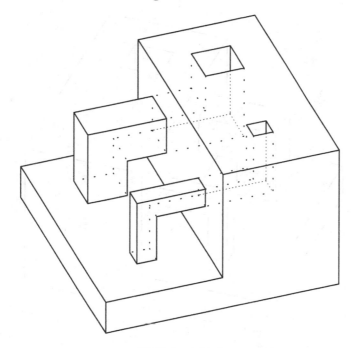

Figure 4.19

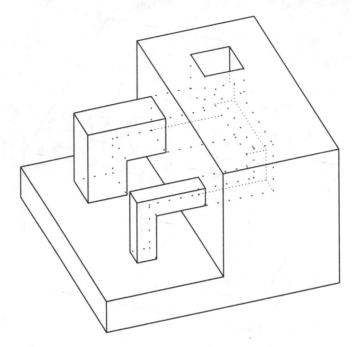

Figure 4.20

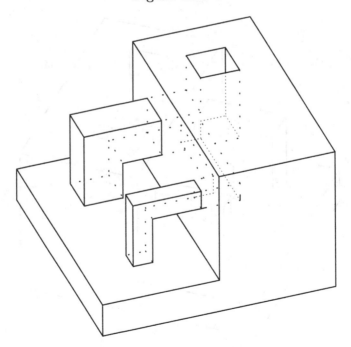

Figure 4.21

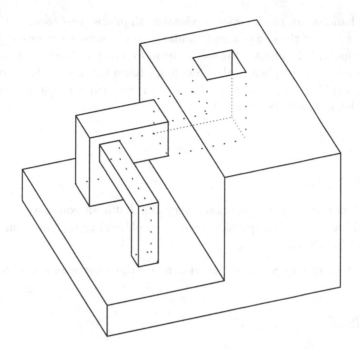

Figure 4.22

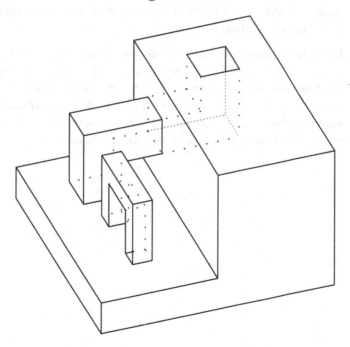

Figure 4.23

Klein handles are much easier to visualize than are cross caps, but we can now accept that, if the end of a real handle is moved across a cross cap instead of a Klein handle, the effect is the same: the moved real handle becomes a Klein handle. To sum up, a sphere with k handles, l Klein handles, and m cross caps is homeomorphic to a sphere with $2k + 2l + m$ cross caps, provided there is at least one Klein handle or cross cap.

Notes

4.1 This innocent-looking operation is very useful: it is an example of something called the *attaching map* and is a good way of building up more complicated spaces from simpler ones.

4.2 Important examples not discussed here are the *orbit spaces*: see [2].

EXERCISES

4.1. Glue twelve thick rods together to form the edges of a cube. The surface of this object is orientable, and so it must be a sphere with a number of handles. How many?

4.2. On a square that represents the torus draw nine paths that join each of three points to each of a separate set of three points: the paths must not cross each other, except that we allow three ends at each of the six points. There are two essentially different solutions to the above problem: give both. Also, give one solution to the problem where the torus is replaced by the Klein bottle.

A polyhedron is a surface constructed from polygons, as the cube is constructed from six squares. We will discuss many well-known polyhedra whose faces are flat regular polygons but first we set polyhedra in a more general topological context, so that polyhedra are not necessarily homeomorphic to the sphere and are built from polygons that are not necessarily flat. In this way, allowing curved faces, we admit many polyhedra, in particular torus-shaped polyhedra, that otherwise would be denied to us. Consideration of these more general polyhedra requires, and motivates, our study of topology.

Our account of topological polyhedra starts by making clear what we mean by a polygon. Given a flat polygon with straight edges we have no problem in deciding how many edges it has but, from a topological point of view, triangles, squares and pentagons are all homeomorphic to a closed disc. How then do we distinguish topologically between polygons with different numbers of edges? The solution is to define a polygon to be not just a set, but a set with additional structure. We define the *standard* n-sided polygon P_n, for each $n \geq 3$, to be the regular polygon lying in the plane that has the origin as centre and $(1, 0)$ as one vertex (or corner), that is, the vertices of P_n are the points $(\cos 2k\pi/n, \sin 2k\pi/n)$ for $k = 1, 2, \ldots, n$. The standard polygons P_3, P_4, P_5, and P_6 are shown in Figure 5.1. The edges and interior of P_n are then determined, where the edges exclude the vertices, and the interior excludes the edges and vertices, although we talk of the vertices at the ends of an edge being *incident with* that edge. An n-sided polygon, or n-gon, is now defined to be a space X, not just homeomorphic to some P_n, but having a specific homeomorphism f from P_n to X associated with it. The edges, vertices, and interior of a

S. Huggett, D. Jordan, *A Topological Aperitif*,
DOI 10.1007/978-1-84800-913-4_5 © Springer-Verlag London Limited 2001, 2009

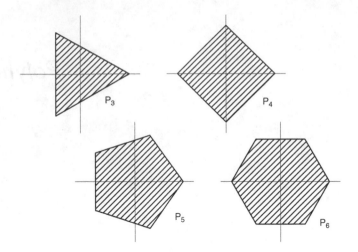

Figure 5.1

polygon X are simply the images under f of the edges, vertices, and interior of P_n and, as with a standard polygon, a vertex may be *incident with* an edge. It is not often in practice that one needs to refer to the homeomorphism f, which is usually kept in the background: only if pressed are we obliged to produce a suitable homeomorphism.

Of course we also use the usual words—triangle, square, pentagon, and so on—for polygons with respectively $3, 4, 5, \ldots$ sides. In our usage a square is a topological square, that is, any polygon with four vertices however stretched or bent. Even though we have a wide interpretation of what a polygon is, it must always be homeomorphic to a disc. We do not accept as polygons either a cylinder formed by bending together and gluing the opposite sides of a square or a triangle bent round so that two vertices coincide.

We give two examples of squares with their homeomorphisms given in detail. An example of a square—still flat here—that looks like a triangle is given by the set T and homeomorphism f of Example 1.2. An example of a bent square is shown in Figure 5.2 and is given by the homeomorphism

$$f(x, y) = (x, y, \sqrt{1 - x^2 - y^2}).$$

A *polyhedron* is a good surface X together with a finite collection of polygons, called the *faces* of the polyhedron, suitably fitting together to make X, that is, satisfying the following requirements. Here, by the edges and vertices of a polyhedron we mean the edges and vertices of its faces.

1. The union of the faces is X.

2. An interior point of any face belongs to no other face.

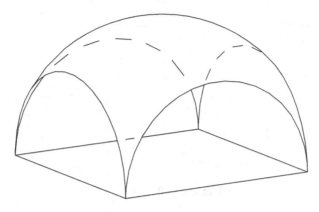

Figure 5.2

3. Each edge of the polyhedron is *incident* with, that is, an edge of, exactly two faces, and no point of the edge belongs to any other face. Two faces having a common edge are called *adjacent*.

4. Each vertex of the polyhedron is *incident* with, that is, a vertex of, at least three edges.

A polyhedron that is homeomorphic to a surface X we refer to as a polyhedron *on* X, for example, a polyhedron on a torus. By a spherical polyhedron we mean a polyhedron on a sphere.

The subtlety of topological polygons can be temporarily left aside while we describe some well-known spherical polyhedra. Our emphasis on these polyhedra, which can perfectly well be discussed without any topological ideas, is justified not only by their inherent appeal but also because they relate to, and aid the understanding of, more general polyhedra.

We shall consider polyhedra from three points of view, more or less simultaneously. Polyhedra can be regarded as metrical entities, using the ideas of lines, planes, distance and angle, or topologically, or as finite, combinatorial entities, consisting only of the incidence relationships between the vertices and the edges and between the edges and the faces.

We start our collection of spherical polyhedra with two infinite families, the prisms and the antiprisms. These polyhedra are made from *regular* polygons, that is, equilateral triangles or other polygons congruent or similar to standard polygons. The *n-prism*, for each $n \geq 3$, consists of two parallel n-sided polygons connected by n squares. As usual, we also refer to a 3-prism as a triangular prism, and so on. All such prisms can be constructed from regular polygons. The 4-prism is, of course, the cube, which is a *regular* polyhedron, the rather overused word "regular" here meaning that the faces of the polyhedron all have

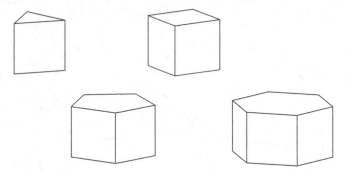

Figure 5.3 Prisms

the same number of edges, and that there are the same number of edges at each vertex: further discussion of ideas of regularity is given later in this chapter. Rather less familiar than the prisms are the *antiprisms,* the *n-antiprism,* for each $n \geq 3$, consisting of two parallel *n*-sided polygons connected by $2n$ triangles. As with the prisms, all the antiprisms can be constructed from regular polygons. The 3-antiprism, or *octahedron,* is regular, but is perhaps more easily perceived as two square pyramids glued together, as in Figure 5.6. The pyramids form a third family of polyhedra, the *n-pyramid,* for each $n \geq 3$, consisting of one *n*-sided polygon, to be thought of as the base, and *n* triangular sides with

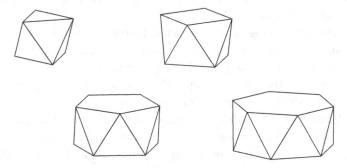

Figure 5.4 Antiprisms

a common vertex. The triangular pyramid, or *tetrahedron,* is regular, and, with the square and pentagonal pyramids, can be made of regular polygons whereas the remaining pyramids cannot. A fourth family then consists of the dipyramids, the *n-dipyramid,* for each $n \geq 3$, being constructed by gluing together two *n*-pyramids by their bases and then disregarding the bases. The *icosahedron* is a regular polyhedron obtained by gluing pentagonal pyramids on a pentagonal antiprism, giving twenty triangles, five at a vertex. The *dodecahedron* has

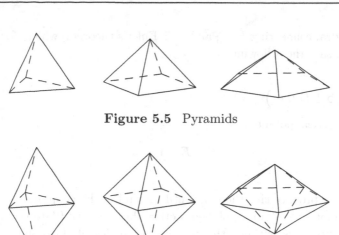

Figure 5.5 Pyramids

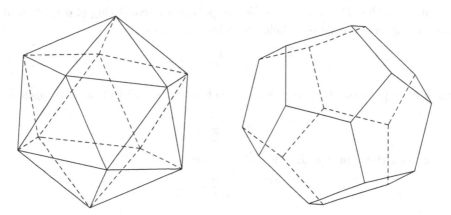

Figure 5.6 Dipyramids

Figure 5.7 Icosahedron and dodecahedron

twelve pentagons, three at a vertex. These five regular spherical polyhedra—
tetrahedron, cube, octahedron, icosahedron, dodecahedron—can all be made
of regular polygons in which form they are the five *Platonic solids.*

To say that polyhedra P and Q are *equivalent* means that there is an *isomorphism* from P to Q, that is, there are bijections from the faces, edges, and
vertices of P to those of Q such that incident pairs are mapped to, and come
from, incident pairs. Sometimes we use the phrase "essentially different" when
we mean not equivalent. By a *symmetry* of P we mean an isomorphism from
P to P. Platonic solids have the expected symmetries.

For the rest of this chapter we shall denote the numbers of faces, edges,
and vertices of a polyhedron by F, E, and V. The all-important central result

for polyhedra, connecting F, E, and V, is Euler's theorem, which, for spherical polyhedra, says the following.

Theorem 5.1 (Euler)

For any spherical polyhedron

$$F - E + V = 2.$$

We defer the proof of this theorem until the end of the chapter.

We cannot have regular polyhedra other than the five Platonic solids made of regular polygons without the sum of the angles of the polygons at each vertex being 2π or more. We now use Euler's theorem to show that there are only the five regular spherical polyhedra even if the faces are topological polygons.

Suppose that P is a regular spherical polyhedron consisting of ϕ-gons, ψ at each vertex. Each edge of P belongs to just two faces, so

$$E = \frac{F\phi}{2}.$$

Each edge joins just two vertices, and each vertex is incident with ψ edges, so

$$E = \frac{V\psi}{2}.$$

By Euler's theorem, $F - E + V = 2$, we have

$$\frac{2E}{\phi} - E + \frac{2E}{\psi} = 2.$$

So

$$\frac{1}{\phi} + \frac{1}{\psi} = \frac{1}{2} + \frac{1}{E} > \frac{1}{2}.$$

The solutions are:

ϕ	ψ	Platonic solid
3	3	Tetrahedron
3	4	Octahedron
3	5	Icosahedron
4	3	Cube
5	3	Dodecahedron

Note that given ϕ and ψ, we can calculate F, E, and V using the equations

$$\frac{1}{\phi} + \frac{1}{\psi} = \frac{1}{2} + \frac{1}{E}, \quad E = \frac{F\phi}{2}, \quad E = \frac{V\psi}{2}.$$

Apart from Euler's theorem, there are other equations or inequalities that are useful. These results are obtained by counting the number of edges in various ways.

Denote by $F_3, F_4, \ldots, F_n, \ldots$ the number of 3-gons, 4-gons, $\ldots$, n-gons, $\ldots$ in a polyhedron. Then if the polyhedron is cut up into its separate faces, the total number of edges is $2E$. However, the number of triangular edges is $3F_3$, the number of square edges is $4F_4$, and so on. Each edge of the polyhedron belongs to just two faces, so

$$3F_3 + 4F_4 + 5F_5 + \cdots = 2E.$$

Because $F_3 + F_4 + F_5 + \cdots = F$, we have

$$3F \leq 2E.$$

If, say, each face has at least four edges, then either directly or from the above equations

$$4F \leq 2E.$$

If, on the other hand, each face has exactly five edges, then either directly or from the above equations

$$5F = 2E.$$

Similarly, there are at least three edges at each vertex, so there are at least $3V$ ends of edges. Thus

$$3V \leq 2E.$$

If every vertex has just three edges, then $3V = 2E$.

Sometimes useful is the following general result connecting vertices and edges. Denote by $V_3, V_4, V_5, \ldots$ the number of vertices incident with $3, 4, 5, \ldots$ edges. Then counting the number of ends of edges gives

$$3V_3 + 4V_4 + 5V_5 + \cdots = 2E.$$

There are other equations that arise from particular conditions. For example, if at each vertex there are ϕ n-sided polygons (and perhaps other types of face), then

$$\phi V = nF_n.$$

Example 5.1

The tetrahedron is a spherical polyhedron having four faces, each pair of faces having a common edge.

Can there be a spherical polyhedron with five or more faces, each pair of faces being adjacent? If so we have a spherical polyhedron where $F \geq 5$ with at least as many edges as there are pairs of faces. Thus $E \geq \frac{1}{2}F(F-1)$. Also $V \leq \frac{2}{3}E$, so that

$$
\begin{aligned}
F - E + V &\leq F - E + \tfrac{2}{3}E \\
&= F - \tfrac{1}{3}E \\
&\leq F - \tfrac{1}{6}F(F-1) \\
&= \tfrac{1}{6}F(7-F) \\
&\leq \tfrac{1}{6}5(7-5),
\end{aligned}
$$

as the parabola $x(7-x)$ has its maximum at $\frac{7}{2}$ and $\frac{7}{2} < 5 \leq F$. So

$$
F - E + V \leq \frac{5}{3} < 2,
$$

which contradicts Euler's theorem, and a spherical polyhedron in which each pair of faces has a common edge must be a tetrahedron.

Example 5.2

Suppose a, b, c, p, q, and r are points on a sphere. Euler's theorem shows the impossibility of the well-known problem of joining each of a, b, c to each of p, q, r, with no paths crossing.

Suppose that this problem had been solved. We must accept here our strong geometric feeling that, as we are on a sphere, such a collection of paths gives rise to a genuine polyhedron (which may not happen on the torus—see Exercise 4.2). Because each path goes from one of a, b, c to one of p, q, r, every polygon must have an even number of edges, and so at least four. Hence $4F \leq 2E$. Also $V = 6$, $E = 9$, so that

$$
F - E + V \leq \frac{1}{2}E - E + 6 = 6 - \frac{1}{2}E = \frac{3}{2} < 2,
$$

contradicting Euler's theorem, which shows that the problem could not have been solved.

The number in Euler's theorem *changes* when we study polyhedra on surfaces other than the sphere, and our next task is to see how.

Polyhedra on a torus

Figure 5.8 is a regular polyhedron on the torus having four squares meeting four at a vertex.

However, in contrast to the sphere, there are infinitely many regular polyhedra on the torus. Figure 5.9 shows some more, all having four squares at each vertex, although, as we shall see shortly in our discussion of regularity, some are more regular than others.

We deduce Euler's theorem for the torus from that for the sphere, in the first step making an appeal to geometric intuition. Let P be a polyhedron on a torus. Cut round the torus in a circle, keeping to edges and vertices of P, so that we obtain a cylinder. Pull the cut ends apart and close them with a new

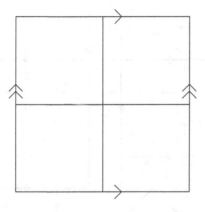

Figure 5.8

polygon each. Suppose that there are n edges round the cut. Then our new, spherical, polyhedron has $F + 2$, $E + n$, and $V + n$ faces, edges, and vertices. But, for our spherical polyhedron $(F + 2) - (E + n) + (V + n) = 2$, so for P we have

$$F - E + V = 0.$$

(If the original cut round the torus has only two edges, then as our faces must have at least three sides, we fill in each end with a pair of triangles instead of a single face: the result follows as before.)

Having seen how Euler's theorem works on the torus, let us analyse the possibilities for regular polyhedra. If P is a regular polyhedron on the torus consisting of ϕ-gons, ψ at a vertex, we now obtain

$$\frac{1}{\phi} + \frac{1}{\psi} = \frac{1}{2} + 0,$$

giving the solutions

$\phi = 3, \psi = 6$
$\phi = 4, \psi = 4$
$\phi = 6, \psi = 3.$

As we have seen in the case $\phi = 4$, the number of faces, edges and vertices is not determined by ϕ, ψ.

We look at these three cases in more detail, but first note that we do not accept the entities shown in Figure 5.10 as polyhedra, as the "squares" and "triangles" are bent round and join up in a complete loop: they are *not* discs.

In fact there are polyhedra on the torus consisting of any number of squares from four up; Figure 5.11 shows one with five squares, and we leave the others as part of Exercise 5.8.

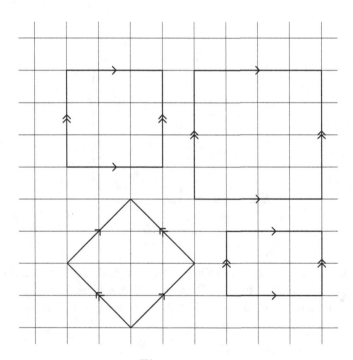

Figure 5.9

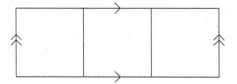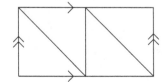

Figure 5.10

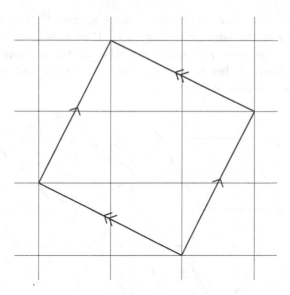

Figure 5.11

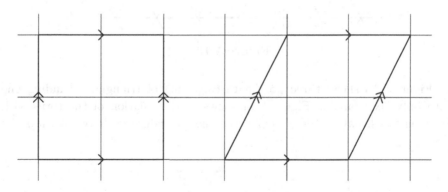

Figure 5.12

Surprisingly, there are two essentially different polyhedra on the torus consisting of four squares, as shown in Figure 5.12. In the first diagram there is a symmetry of the polyhedron sending an edge to any other edge: in the second there is not, which we now explain in more detail. In the first diagram we can attach the number 2 to each edge, as follows. Starting at a given edge, choose one of the incident squares, and move to its opposite edge. Repeat this process, without retracing your steps, until you return to the original given edge. In this case we get two distinct edges. An alternative description of this process is to find a sequence of distinct edges $e_1, e_2, \ldots, e_n$ such that each pair of edges

$e_1e_2, e_2e_3, \ldots, e_{n-1}e_n, e_ne_1$ is opposite in some square. We call the number n the *ring* number of the edge. (See Note 5.1.) Here n is 2. In the second diagram, vertical edges have ring number 2, whereas the horizontal edges have ring number 4. There is therefore no symmetry of the polyhedron that sends a vertical edge to a horizontal edge.

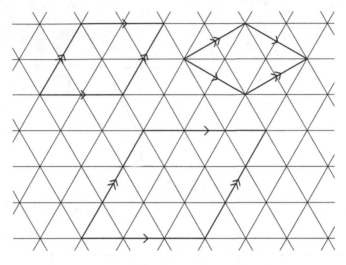

Figure 5.13

Figure 5.13 shows three polyhedra consisting of triangles, including one with only six triangles. Figure 5.14 shows a triangulation of the torus with fourteen faces, having the property that no two edges go between the same

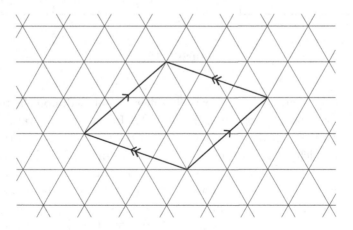

Figure 5.14

pair of vertices. Let us now use Euler's theorem to show that fourteen is the smallest number for such a triangulation. Euler's theorem gives

$$F - E + V = 0.$$

As the polyhedron consists of triangles we have

$$3F = 2E,$$

and as there are at least as many pairs of vertices as edges

$$E \le \frac{1}{2}V(V - 1).$$

Hence $E = \frac{3}{2}F$, $V = \frac{1}{2}F$, and $F \ge 14$.

Figure 5.15 shows a polyhedron on the torus consisting of three hexagons. All three hexagons have the same six vertices, and any two hexagons have three common edges. Figure 5.16 shows two polyhedra each consisting of seven hexagons. The first has the property that each pair of hexagons is adjacent, so that any colouring of the hexagons, with adjacent hexagons having different colours, requires seven colours. (See Note 5.2.) The second polyhedron, on the other hand, requires only four colours, so these two polyhedra must be essentially different. The concept of ring number also applies to hexagonal polyhedra, because all we really used was the idea of opposite edges in a face. Both of the polyhedra of Figure 5.16 have ring number 7 for all edges, so ring number does not distinguish them.

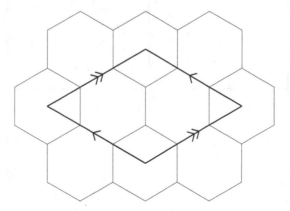

Figure 5.15

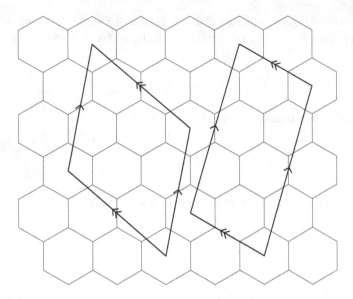

Figure 5.16

Polyhedra on the Klein bottle

Our discussion of the Euler formula for the torus holds equally well for the Klein bottle, and so again

$$F - E + V = 0.$$

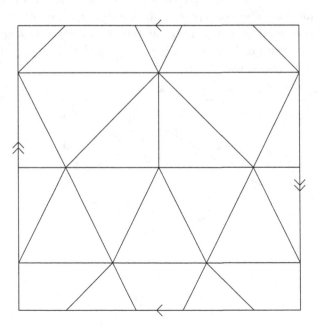

Figure 5.17

Therefore, we have the same three cases for ϕ and ψ, and indeed many of our diagrams of polyhedra on the torus lead to polyhedra on the Klein bottle simply by reversing one of the arrows. One has to check (even for the torus) that the identifications do give a polyhedron: reversing an arrow in the identifications of the polyhedra in Figures 5.11 and 5.14 for example, do not.

Because the Euler formula has not changed, it is still true that a triangulation of the Klein bottle having the property that no two edges go between the same pair of vertices, must have at least fourteen faces. We do not believe that there is one with fourteen, but Figure 5.17 shows one with sixteen.

Polyhedra on the real projective plane

Because the real projective plane is homeomorphic to the sphere with antipodal points identified, given any polyhedron P on real projective plane we can find a spherical polyhedron having twice as many faces, edges, and vertices as P has. Thus, if F, E, and V refer to P, then $2F - 2E + 2V = 2$ and so $F - E + V = 1$. Any polyhedron on S^2 such that each antipodal pair of points is either a pair of vertices, or a pair of edge points, or a pair of interior points gives rise to a polyhedron on the real projective plane.

For example, a cube gives a regular polyhedron on the real projective plane with $\phi = 4$, $\psi = 3$, $F = 3$, $V = 4$, and $E = 6$, and is shown in Figure 5.18.

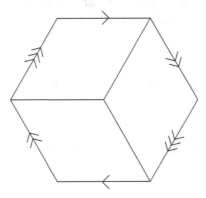

Figure 5.18

This is another example of a polyhedron with only three faces.

In fact, if X is any good surface, $F - E + V$ is the same for all polyhedra on X, this common number being called the *Euler number,* or Euler characteristic, of X, and denoted by χ. We give a brief, intuitively convincing, argument that each good surface has an Euler number.

Sphere with k handles, the surface of genus k

To show that a sphere with k handles has an Euler number we can use exactly
the same method as for the torus, but k times: the torus is merely a sphere
with one handle, after all. So we make k cuts, one in each handle, and fill in the
cut ends. If we do this one cut at a time, we can see from the torus argument
that each such step increases χ by 2, and we must eventually obtain a sphere,
which has $\chi = 2$. So the Euler number of the original space must have been
$2 - 2k$.

Here is an example of a regular polyhedron on a double torus. Take two
copies of the polyhedron on the torus shown in Figure 5.8, glue the two polyhedra together, a square from one to a square from the other. Remove the glued
pair and we have a polyhedron, on a double torus, consisting of six squares. All
six squares have the same four vertices.

Sphere with n cross caps

To show that a sphere with n cross caps has an Euler number we regard the
surface as a sphere with $n - 1$ handles with opposite points identified. We
use the same argument as in the case of the real projective plane: let P be a
polyhedron (with F, E, V faces, edges, and vertices) on a sphere with n cross
caps. There must be a polyhedron (with $2F, 2E, 2V$ faces, edges, and vertices)
on the sphere with $n - 1$ handles, and hence

$$2F - 2E + 2V = 2 - 2(n - 1).$$

So the Euler number of a sphere with n cross caps is $2 - n$.

The Classification Theorem

We summarise these results in the important theorem below, which states how
the orientability (or otherwise) and Euler number of a surface completely *de-
termine* it. (The word "compact" in the theorem is the same as closed and
bounded in the case when X is a Euclidean set, as we show in Appendix A.)

Theorem 5.2

Suppose that the surface X is compact, path-connected, and has Euler number χ. Then, if X is not orientable, it is homeomorphic to a sphere with $2 - \chi$ cross caps. If X is orientable, then χ is even and X is homeomorphic to a sphere with $\frac{1}{2}(2 - \chi)$ handles.

Dual polyhedra

Let P be a polyhedron. The construction of a *dual* Q of P, illustrated in Figure 5.19, starts by choosing, for each face A of P, a vertex v_A in the interior of A. Let e be an edge of P incident with A, B. Draw a line between the corresponding vertices v_A, v_B of Q, going through A, B and some point of e, and not crossing itself. This can be done for all edges of P so that no two lines meet except at the vertices of Q. The set obtained by removing these lines from P has components that form the interiors of the faces of Q. Each face of Q contains just one vertex of P.

The construction of Q automatically gives bijections from the faces, edges, and vertices of P to the vertices, edges, and faces of Q respectively in such a way that an incident pair of P is sent to an incident pair of Q. Thus Q is the same as P from an incidence point of view, except that what were faces are now vertices, and vice versa.

If P has F faces, E edges, and V vertices, then Q has V faces, E edges, and F vertices. If Q' is any dual of P, then Q' is equivalent to Q, and we therefore talk of *the* dual of Q. The word dual is used because, if Q is a dual of P, then P is a dual of Q.

The cube and octahedron form a dual pair, as do the dodecahedron and icosahedron. The tetrahedron is self-dual. Every pyramid is self-dual. The dual of a prism is a dipyramid.

Regularity

The weak idea of regularity used so far, namely, that all faces have the same number of edges and all vertices have the same number of edges, is a *local* idea, that is, it can be determined by looking at vertices and faces individually. A better interpretation of our feeling of regularity is a *global* idea, that is, the whole polyhedron appears the same when viewed in various ways.

The strongest possible form of regularity for a polyhedron P, which we call *fully symmetric*, is that given vertices v_1, v_2, edges e_1, e_2, and faces F_1, F_2, where v_1 is incident with e_1, which is incident with F_1, and v_2 is incident with e_2, which is incident with F_2, there is a symmetry of P that sends v_1, e_1, F_1 respectively to v_2, e_2, F_2. Because each edge of a polyhedron is incident with

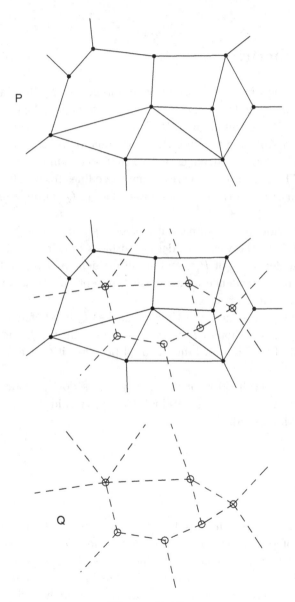

Figure 5.19

two vertices and two faces, being fully symmetric is equivalent to saying that a polyhedron has $4E$ symmetries. The five Platonic solids are fully symmetric and, indeed, each symmetry of these polyhedra can be realised by an isometry of space, in fact a rotation or reflection, that produces the given symmetry. Of those polyhedra on the torus shown in Figures 5.8 and 5.9 only the polyhedron with six squares fails to be fully symmetric—both 2 and 3 occur as ring numbers and there is no symmetry sending a vertical edge to a horizontal edge.

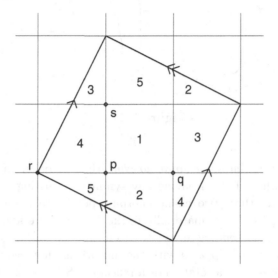

Figure 5.20

A polyhedron can be regular in our first sense and have a single ring number and yet not be fully symmetric. In the polyhedron P shown in Figure 5.20 every edge has ring number 5. Now suppose that there is a symmetry of P which leaves the vertices p, q and the edge pq fixed but interchanges faces 1 and 2—a reflection in pq. Considering the edge pr we see that such a symmetry must interchange faces 4 and 5, but considering instead the edge qs the symmetry must interchange faces 4 and 3. Hence the polyhedron is not fully symmetric. A similar argument applies to the polyhedra of Figure 5.16.

For a *semi-regular* polyhedron P we relax the condition, required of regular polyhedra, that all faces have the same number of edges, but we do require that the polyhedron looks the same from each vertex. That is, given vertices v_1, v_2 there is a symmetry of P sending v_1 to v_2. There are thirteen semi-regular spherical polyhedra apart from the prisms and antiprisms, called Archimedean solids, and all of these can be made with regular polygons. Three of the thirteen are shown in Figure 5.21.

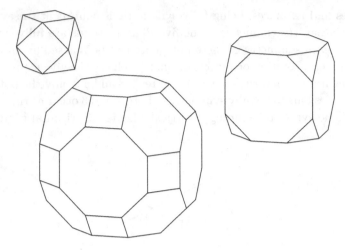

Figure 5.21

Two of the thirteen are *enantiomorphic,* that is, occur in left and right hand forms, neither of them having any symmetries arising from reflections. The distinction of these two enantiomorphic polyhedra among the thirteen manifests itself purely combinatorially in the fact that the identity symmetry is the only symmetry sending any vertex to itself.

A new phenomenon appears with the antiprisms: not every symmetry of an antiprism arises from a rotation or a reflection. Some symmetries require a rotation followed by a reflection in a plane perpendicular to the axis of rotation.

Semi-regular polyhedra give a very good example of the importance of distinguishing the global from the local: there is a fourteenth, bogus, Archimedean polyhedron made of regular polygons in which all the vertices are locally the same. The real Archimedean and its bogus counterpart are shown in Figure 5.22.

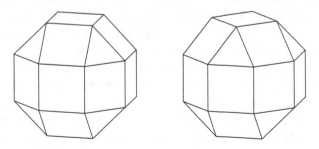

Figure 5.22

Semi-regular polyhedra on the torus can be constructed by identifying opposite sides of a parallelogram in one of the eight semi-regular plane tessellations.

Example 5.3

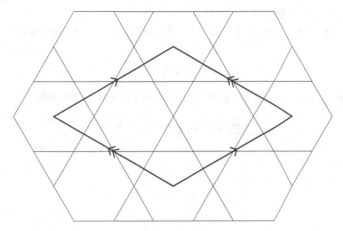

Figure 5.23

Figure 5.23 shows a semi-regular polyhedron on the torus consisting of three hexagons and six triangles. This polyhedron, or rather its dual, shows that on the torus three points can each be joined to each of a (separate) set of six points, no two paths crossing. Take the three points to be the dual vertices in the three hexagons and the six points to be the dual vertices in the six triangles; the dual edges provide the required paths.

Proof of Euler's Theorem

We give a proof of Euler's theorem, due to von Staudt (1847), usually called the tree proof.

Let P be a spherical polyhedron. Because the sphere is path-connected, any two given vertices are connected by a path. This path can be chosen to go along edges and through vertices, though because edges can be pathologically twisted this is fairly hard to show, and we omit the argument.

Find a tree T in P that consists of all the vertices of P and some, E_1 say, of the edges. Then $V = E_1 + 1$. Now construct a graph G consisting of the vertices of the polyhedron *dual* to P and the edges of the dual corresponding to those E_2 edges of P *not* among the E_1 edges of T.

   The crucial point of the proof is that G is also a tree. To show this we have to show two things, that G is connected and that it contains no circuit. Firstly, if G were not connected then it is intuitively clear that parts of G would be separated by a circuit of T. Secondly, if G contained a circuit then that circuit would disconnect T (see Note 5.3).

   But G has F vertices, so $F = E_2+1$. Since $E_1+E_2 = E$, our three equations now give

$$F - E + V = 2.$$

In fact G is connected whether P is spherical or not, so always

$$F \leq E_2 + 1 \text{ and } F - E + V \leq 2.$$

This completes the proof.

Notes

5.1 This definition is analogous to that of the *order* of an element of a group.

5.2 In fact, seven is the largest number ever needed for colouring *any* polyhedron on the torus.

5.3 This amounts to the famous and deep *Jordan curve theorem*, proved by *Camille* Jordan in 1887, see [2].

EXERCISES

5.1. The spherical polyhedron P has at least five edges at each vertex. Show that P contains at least twenty triangles.

5.2. Show that any spherical polyhedron having a square and two hexagons at each vertex must consist of six squares and eight hexagons. Describe how such a polyhedron can be constructed from an octahedron.

5.3. At each vertex of the spherical polyhedron P there are just three faces, one square, one hexagon, and one decagon. How many vertices does P have? How many edges does P have? How many squares, hexagons, and decagons does it have?

5.4. At each vertex of the spherical polyhedron P there are either four squares, or two squares and a pentagon. Show that P has just two pentagons. Give two examples of such a polyhedron.

5.5. Show that there is a spherical polyhedron that has two edges joining the same pair of vertices.

5.6. Show that no spherical polyhedron can consist of just three faces.

5.7. (a) Show that a polyhedron on the torus with no triangles, squares, or pentagons must consist entirely of hexagons.

   (b) Deduce that it is impossible to have a polyhedron on the torus consisting of eight faces, each pair of faces being adjacent.

5.8. (a) Show that there is a polyhedron on the torus consisting of n squares for each $n \geq 4$.

   (b) Show that there is a polyhedron on the torus consisting of n hexagons for each $n \geq 3$.

   (c) Show that there is a polyhedron on the torus consisting of $2n$ triangles for each $n \geq 3$.

5.9. Let P be a spherical polyhedron consisting of triangles and hexagons. Show that P has at least four triangles. Give two examples of such a polyhedron, one example having four hexagons and the other having just two hexagons. Give two examples of a polyhedron on the torus, one example consisting of four hexagons and eight triangles, and the other consisting of five hexagons and ten triangles.

5.10. The spherical polyhedron P consists of triangles and squares and has exactly four edges at each vertex. Show that P contains just eight triangles. Find four essentially different such polyhedra P. Give an example of a polyhedron on the torus that consists of triangles and squares and has exactly five edges at each vertex.

5.11. Show that no polyhedron P consisting of squares, six at each vertex, can exist on a sphere or a torus.

5.12. Show that the result of Example 5.3 is best possible in the following sense: show that no polyhedron on the torus can have ten vertices, split into sets of three and seven, with edges joining each of the three to each of the seven.

Show that, on the torus, it is possible to join each of four points to each of a separate set of four points, no two paths crossing. Show that, in the sense above, four is the best possible result.

5.13. Construct a (part of an infinite) "polyhedron" in real space and made of real squares, six at each vertex. This polyhedron is more regular than a cube in that it is fully symmetric and the "inside" and "outside" are congruent.

6
Winding Number

We hope that at this stage the reader has some feel for surfaces and their classification by orientability and Euler number. Our discussions of orientability and Euler number have rested fundamentally on appeal to geometric intuition and, however convincing, cannot be accepted as rigorous proof. We *can* prove some results in this direction, though, by using the idea of winding number. In particular we can distinguish the sphere from other good surfaces and justify the distinction between edge and interior points, so that for example a closed Möbius band, with path-connected edge, cannot be homeomorphic to a cylinder.

By a *closed path* in the space S we mean a continuous mapping $\gamma : [a, b] \to S$ where $a < b$ and $\gamma(a) = \gamma(b)$. Let γ be a closed path in the plane not passing through the point $\mathbf{p}$. Then the *winding number* $W(\gamma, \mathbf{p})$ of γ about $\mathbf{p}$ means the number of times γ goes round $\mathbf{p}$ in an anticlockwise sense.

Figure 6.1 indicates a path γ where we have $W(\gamma, \mathbf{p}) = 0$, $W(\gamma, \mathbf{q}) = 1$, $W(\gamma, \mathbf{r}) = 2$, and $W(\gamma, \mathbf{s}) = -1$. Although the idea of winding number is intuitively very clear we need to articulate the idea sharply if it is to provide the required proofs. Each point $\mathbf{x}$ in the plane can be written as $|\mathbf{x}|(\cos\theta, \sin\theta)$ where, for $\mathbf{x}$ other than the origin, θ is determined up to a multiple of 2π. Let $\gamma : [a, b] \to \mathbb{R}^2$ be a closed path not passing through the origin. Then for each $t \in [a, b]$ there is a $\theta(t)$ such that

$$\gamma(t) = ||\gamma(t)||(\cos\theta(t), \sin\theta(t)).$$

Crucially, we can find a *continuous* function θ satisfying the above equation for all $t \in [a, b]$. (A brief proof can be found in Appendix A.)

S. Huggett, D. Jordan, *A Topological Aperitif*,
DOI 10.1007/978-1-84800-913-4_6 © Springer-Verlag London Limited 2001, 2009

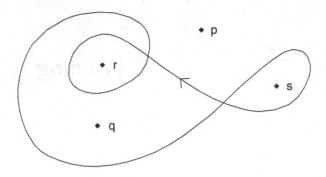

Figure 6.1

Similarly, if γ does not pass through the point $\mathbf{p}$, there is a continuous function θ such that

$$\gamma(t) = \mathbf{p} + ||\gamma(t)||(\cos\theta(t), \sin\theta(t)).$$

Such a continuous function we call an *angle function* for γ about $\mathbf{p}$. If θ and ϕ are both angle functions for γ about $\mathbf{p}$ then, for each $t \in [a, b]$,

$$\phi(t) - \theta(t) = 2\pi k_t$$

for some integer k_t. Also, as θ and ϕ are continuous, $\phi - \theta$ is continuous. Consequently $\phi - \theta$ is constant, so that $\phi(b) - \phi(a) = \theta(b) - \theta(a)$ and we may make the following definition.

Definition 6.1

Let $\gamma : [a, b] \to \mathbb{R}^2$ be a closed path not passing through the point $\mathbf{p}$. Then $W(\gamma, \mathbf{p})$, the *winding number* of γ about $\mathbf{p}$, is

$$\frac{\theta(b) - \theta(a)}{2\pi},$$

where θ is any angle function for γ about $\mathbf{p}$.

Example 6.1

Let $\gamma_n(t) = R(\cos nt, \sin nt)$, where $t \in [0, 2\pi]$, $R > 0$, and n is an integer. Then we may take nt as an angle function for γ_n about the origin, so that

$$W(\gamma_n, (0, 0)) = (n2\pi - 0)/2\pi = n.$$

Example 6.2

Let $\mathbf{p}$ and $\mathbf{q}$ be distinct points in the plane and put $\gamma(t) = \mathbf{p}$ for all $t \in [a, b]$. Then any angle function for γ about $\mathbf{q}$ is constant, and $W(\gamma, \mathbf{q}) = 0$.

Except when γ is very simple, as in the above examples, we do not calculate winding number directly, but can often make use of the following results, the first being essential for later use.

Theorem 6.1

Suppose that γ and δ are closed paths in the plane, neither passing through the point $\mathbf{p}$ and both having domain $[a, b]$. Suppose also that $\gamma(t)$ and $\delta(t)$ are never on directly opposite rays from $\mathbf{p}$. Then $W(\gamma, \mathbf{p}) = W(\delta, \mathbf{p})$.

Proof

Let θ and ϕ be angle functions for γ and δ about $\mathbf{p}$. As $\gamma(t)$ and $\delta(t)$ are never on directly opposite rays from $\mathbf{p}$ it follows that $\theta(t) - \phi(t)$ is never an odd multiple of π. Now $(\theta - \phi)(b) - (\theta - \phi)(a) = 2\pi m$ for some integer m, which, as $\theta - \phi$ is continuous and never an odd multiple of π, must be zero or the intermediate value theorem would be violated. (A continuous function taking values y_1 and y_2 takes all values y in between y_1 and y_2.) Hence $\theta(b) - \theta(a) = \phi(b) - \phi(a)$ and $W(\gamma, \mathbf{p}) = W(\delta, \mathbf{p})$. This completes the proof.

Example 6.3

Let $\gamma : [0, 8] \to \mathbb{R}^2$ be a path that goes with unit speed anticlockwise round a square of side 2, starting at $(1, 0)$, as in Figure 6.2. Put

$$\delta(t) = \left(\cos \frac{2\pi t}{8}, \sin \frac{2\pi t}{8} \right),$$

for $t \in [0, 8]$. Then $\gamma(t)$ and $\delta(t)$ are never on directly opposite rays from the origin, and

$$W(\gamma, (0, 0)) = W(\delta, (0, 0)) = 1.$$

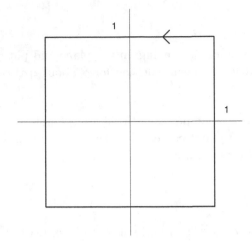

Figure 6.2 Example 6.3

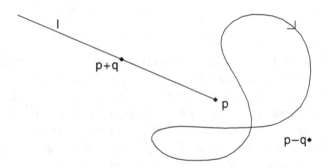

Figure 6.3 A ray not meeting the path

Theorem 6.2

Suppose that $\mathbf{p}$ and $\mathbf{q}$ are points in the plane, $\mathbf{q} \neq (0,0)$, and that l is the ray $\{\mathbf{p} + k\mathbf{q} : k \geq 0\}$. Suppose that the closed path γ does not meet l. Then $W(\gamma, \mathbf{p}) = 0$.

Proof

Let δ be the closed path given by $\delta(t) = \mathbf{p} - \mathbf{q}$, where δ and γ have the same domain. Then $\gamma(t)$ and $\delta(t)$ are never on directly opposite rays from $\mathbf{p}$, and

$$W(\gamma, \mathbf{p}) = W(\delta, \mathbf{p}) = 0.$$

This completes the proof.

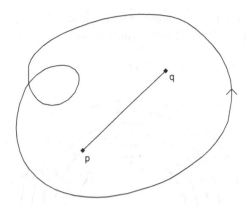

Figure 6.4 A segment not meeting the path

Theorem 6.3

Let **p** and **q** be points in the plane and suppose that the closed path $\gamma : [a, b] \to \mathbb{R}^2$ does not meet the line segment joining **p** to **q**. Then $W(\gamma, \mathbf{q}) = W(\gamma, \mathbf{p})$.

Proof

Let θ and ϕ be angle functions for γ about **p** and **q**. As $\gamma(t)$ is never on the line segment joining **p** to **q**, $\theta(t) - \phi(t)$ can never be an odd multiple of π so, as in the proof of Theorem 6.1 we have $W(\gamma, \mathbf{q}) = W(\gamma, \mathbf{p})$. This completes the proof.

Example 6.4

Let C be the circle $C(t) = \mathbf{p} + R(\cos t, \sin t)$, $t \in [0, 2\pi]$, where $R > 0$. Direct calculation shows that $W(C, \mathbf{p}) = 1$ and Theorem 6.3 tells us that $W(C, \mathbf{q}) = 1$ whenever **q** is inside C. Theorem 6.2 tells us that $W(C, \mathbf{q}) = 0$ whenever **q** is outside C.

On its own the idea of winding number achieves little, but combined with the idea of deformation it gives valuable results. In this book we restrict ourselves to deformation of closed curves and do not discuss deformation of surfaces.

Definition 6.2

Let γ and δ be closed paths in a space S, both paths having domain $[a, b]$. A *deformation (or homotopy)* in S from γ to δ consists of a closed path γ_u in S

with domain $[a, b]$ for each u in some interval $[c, d]$. We require $\gamma_c = \gamma$, $\gamma_d = \delta$ and the mapping $(u, t) \rightarrow \gamma_u(t)$ to be continuous.

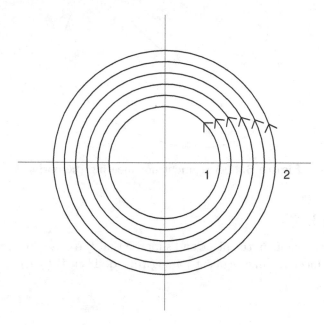

Figure 6.5   Example 6.5

Example 6.5

Put $\gamma_u(t) = u(\cos t, \sin t)$, $t \in [0, 2\pi]$, $u \in [1, 2]$. Then γ_u is a deformation in the plane from γ_1 to γ_2, and is indicated in Figure 6.5. In fact γ_u can also be regarded as a deformation in the plane with the origin removed.

Example 6.6

Let $\gamma : [a, b] \rightarrow \mathbb{R}^2$ be any closed path in the plane. We can shrink γ to the origin with the deformation

$$\gamma_u(t) = (1 - u)\gamma(t), \, t \in [a, b], \, u \in [0, 1].$$

The same argument applies to any path in the unit disc.

Example 6.7

We shrink the equator of S^2 to the north pole, keeping the deformation in S^2 : put

$$\gamma_u(t) = (\sqrt{1-u^2}\cos t, \sqrt{1-u^2}\sin t, u),\ t \in [0, 2\pi],\ u \in [0,1].$$

Example 6.8

Returning to the notation of Example 3.6, let γ be the $(0,1)$-circle given by $\gamma(t) = \tau(0,t)$ for $t \in [0,1]$, and let δ be the opposite circle, given by $\delta(t) = \tau(1/2, t)$ for $t \in [0,1]$. Then γ is deformed into δ by

$$\gamma_u(t) = \tau(u,t), \quad t \in [0,1], \quad u \in [0, 1/2].$$

Theorem 6.4

Suppose that $\gamma, \delta : [a,b] \to S$ are closed paths in S and $\gamma_u,\ u \in [c,d]$ is a deformation in S from γ to δ. Suppose also that $f : S \to T$ is continuous. Then $f\gamma$ and $f\delta$ are are closed paths in T and $f\gamma$ can be deformed in T to $f\delta$.

Proof

The deformation $f\gamma_u,\ u \in [c,d]$ deforms $f\gamma$ to $f\delta$. This completes the proof.

Example 6.9

Let p be any point of S^2. Rotating and then projecting stereographically gives a homeomorphism f from $S^2 \setminus \{p\}$ to the plane. Let γ be any closed path in $S^2 \setminus \{p\}$. Then $f\gamma$ is a closed path in $\mathbf{R}^2$, which, by Example 6.6 above, can be shrunk to the origin, using a deformation γ_u say. Using Theorem 6.4, the deformation $f^{-1}\gamma_u$ in $S^2 \setminus \{p\}$ deforms $f^{-1}f\gamma$, which is γ, to a point. Thus every closed path in $S^2 \setminus \{p\}$ can be shrunk to a point in S^2, the deformation being in $S^2 \setminus \{p\}$. In fact, every closed path in S^2 can be shrunk to a point in S^2, the deformation being in S^2, but this is harder to prove, a difficulty being that there are pathological closed paths in S^2 that pass through every point of S^2.

We now come to the essential connection between winding number and deformation, namely that winding number is invariant, that is, stays the same, under deformation.

Theorem 6.5

Let $\gamma_u,\ u \in [c,d]$, be a deformation in $\mathbf{R}^2 \setminus \{(0,0)\}$ from γ to δ. Then

$$W(\gamma, (0,0)) = W(\delta, (0,0)).$$

Proof

First suppose that each γ_u lies in the unit circle. As $W(\gamma_u, (0,0))$ is always an integer the result follows if we show that $u \to W(\gamma_u, (0,0))$ is continuous. We do this by showing that, for each v in $[c,d]$, there is a neighbourhood of v in which $W(\gamma_u, (0,0))$ is constant.

Suppose that $W(\gamma_u, (0,0))$ is not constant in any neighbourhood of v. Then, for $n = 1, 2, \ldots$ there is some point u_n in $[c,d]$, within distance $\frac{1}{n}$ of v, such that

$$W(\gamma_{u_n}, (0,0)) \neq W(\gamma_v, (0,0)).$$

Hence there is some t_n in $[a,b]$, the domain of each γ_u, such that $\gamma_{u_n}(t_n)$ and $\gamma_v(t_n)$ are on opposite rays from the origin, which here means diametrically opposite points of the unit circle. Because each t_n is in $[a,b]$, there is a point w in $[a,b]$ such that every neighbourhood of w contains some t_n. Every neighbourhood of (v,w) contains both (u_n, t_n) and (v, t_n) for some n. But $\gamma_{u_n}(t_n)$ and $\gamma_v(t_n)$ are diametrically opposite points of S^1. This shows that $(u, t) \to \gamma_u(t)$ is not continuous at (v, w), a contradiction.

To prove the result for general γ_u let $f : \mathbb{R}^2 \setminus \{(0,0)\} \to S^1$ be the mapping

$$\mathbf{x} \to \frac{\mathbf{x}}{\|\mathbf{x}\|}.$$

Then f is continuous. So $f\gamma_u$ is a deformation in S^1 of $f\gamma$ to $f\delta$ and $W(f\gamma, (0,0)) = W(f\delta, (0,0))$. But an angle function for γ about $(0,0)$ is an angle function for $f\gamma$ about $(0,0)$, so $W(f\gamma, (0,0)) = W(\gamma, (0,0))$. Similarly $W(f\delta, (0,0)) = W(\delta, (0,0))$. Consequently

$$W(\gamma, (0,0)) = W(\delta, (0,0)).$$

This completes the proof.

Example 6.10

No two of the paths γ_n from Example 6.1 can be deformed into each other in $\mathbb{R}^2 \setminus \{(0,0)\}$, as $W(\gamma_n, (0,0)) = n$.

Example 6.11

Put $\gamma(t) = (\cos t, \sin t, h)$, $t \in [0, 2\pi]$ and let C be the cylinder $\{(x, y, z) : x^2 + y^2 = 1\}$. We show that, in C, the closed path γ cannot be shrunk to a point. Suppose that γ_u, $u \in [c, d]$ shrinks γ to a point, the deformation being in C. Let f be the projection $(x, y, z) \to (x, y)$.

Then $f\gamma_u$ shrinks $f\gamma$ to a point, the deformation being in S^1. But $f\gamma$ is the path $t \to (\cos t, \sin t)$, $t \in [0, 2\pi]$, which has winding number 1 about the origin, whereas a point has zero winding number. Thus no such deformation γ_u is possible. Hence C is not homeomorphic to the plane or a disc. Letting q be any point of C not on γ, we see that there is a closed path in $C \setminus \{q\}$ that cannot be shrunk to a point, the deformation being in $C \setminus \{q\}$. So, from Example 6.9, the cylinder is not homeomorphic to the sphere. Similarly, the closed cylinder $\{(x, y, z) : x^2 + y^2 = 1, |z| \leq 1\}$ is not homeomorphic to S^2 or a disc.

Example 6.12

An argument similar to that above applies to the torus T_S. We use the notation of Example 3.6. The projection $f(x, y, z) = (x, y)$ sends no point of T_S to the origin, so a deformation in T_S projects to a deformation in $\mathbb{R}^2 \setminus \{(0, 0)\}$. The $(1, 0)$-circle in T_S given by $\gamma(t) = \tau(t, 0)$ for $t \in [0, 1]$ projects to a circle in $\mathbb{R}^2 \setminus \{(0, 0)\}$ that has angle function $\theta(t) = 2\pi t$, and so winding number 1 about the origin. Hence T_S contains a closed path that cannot be shrunk to a point in T_S, the deformation being in T_S. Arguing as in the above example it now follows that the torus is not homeomorphic to the sphere.

The same idea can be used to prove that the sphere is not homeomorphic to a sphere with k handles for any $k \geq 1$.

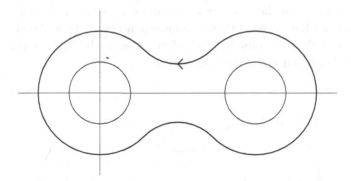

Figure 6.6

Such a sphere with handles can be positioned in space so that no point projects (under $f(x, y, z) = (x, y)$) to the origin, and there is a closed path in the surface that projects to a closed path in $\mathbb{R}^2 \setminus \{(0, 0)\}$ having winding number 1.

Example 6.13

We prove, without appealing to orientability, that the sphere is not homeomorphic to the real projective plane, which we regard here as S^2 with opposite points identified.

First, let $g : \mathbb{R}^2 \to \mathbb{R}^2$ be $z \to z^2$, where the point z in the plane is thought of as a complex number. Let the projection $f : \mathbb{R}^3 \to \mathbb{R}^2$ be as before, namely $(x, y, z) \to (x, y)$, and put $h(x, y, z) = g(f(x, y, z))$. Then

$$h(x, y, z) = h(-x, -y, -z)$$

and there is a continuous mapping $\hat{h}$ (see Appendix A) from the real projective plane to $\mathbb{R}^2$ that sends the pair $\{(x, y, z), (-x, -y, -z)\}$ to $h(x, y, z)$.

We define a closed path γ in the real projective plane by taking $\gamma(t)$ to be the opposite pair containing $(\cos t, \sin t, 0)$ for $t \in [0, \pi]$: note that γ *is* a closed path. Now $\hat{h}$ sends the real projective plane with the pair $\{N, S\}$ of poles removed to $\mathbb{R}^2 \setminus \{(0, 0)\}$, and $\hat{h}\gamma$ is the closed path $t \to (\cos 2t, \sin 2t)$, $t \in [0, \pi]$, which has winding number 1 about the origin. Hence $\hat{h}\gamma$ cannot be shrunk to a point by a deformation in $\mathbb{R}^2 \setminus \{(0, 0)\}$, and γ cannot be shrunk to a point by a deformation in the real projective plane with $\{N, S\}$ removed. Hence the sphere is not homeomorphic to the real projective plane.

The real projective plane with $\{N, S\}$ removed is homeomorphic to the real projective plane with the Arctic and Antarctic removed, and so is a Möbius band. Hence we have proved that a Möbius band contains a closed path that cannot be shrunk to a point by any deformation in the Möbius band. Hence the Möbius band is not homeomorphic to a disc.

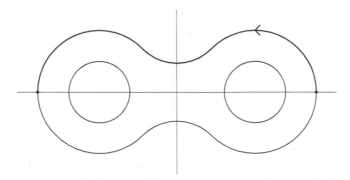

Figure 6.7

We adapt the above argument for the real projective plane to show that the sphere is not homeomorphic to a sphere with n cross caps, for $n \geq 1$. Take a sphere with k handles positioned in space as in Figure 6.7 so that opposite points can be identified to give a sphere with $k+1$ cross caps. Remove the pair of points above and below the origin as before and replace the closed path γ previously used by the closed path indicated.

Example 6.14

We are now in a position to prove that no two different (m, n)-circles in the torus can be deformed into each other, the deformation being in the torus. In this example we take the torus to be T^2 and regard the (m, n)-circle as being the closed path given by

$$\gamma(t) = (\cos 2\pi mt, \sin 2\pi mt, \cos 2\pi nt, \sin 2\pi nt), \; t \in [0, 1].$$

Let g be the projection $(x, y, z, w) \rightarrow (x, y)$. Then

$$g\gamma(t) = (\cos 2\pi mt, \sin 2\pi mt), \; t \in [0, 1],$$

and $g\gamma$ has winding number m about the origin. Moreover g projects the whole torus to S^1. So (m, n)-circles for different m cannot be deformed into each other. Similarly, using the projection $h(x, y, z, w) = (z, w)$ we see that (m, n)-circles for different n cannot be deformed into each other. Hence no two different (m, n)-circles can be deformed into each other in the torus.

We can use winding number to prove that there is a valid distinction between edge points and interior points of a surface with boundary, the idea being that an interior point has a closed path going round it whereas an edge point does not.

Let p be a point of a surface with boundary that has a neighbourhood mapped to the semicircle $H = \{(x, y) : x^2 + y^2 < 1, \; x \geq 0\}$, the point p being mapped to the origin. We call such a point an *edge point* of the surface. Given such an edge point, *any* neighbourhood of p contains a neighbourhood N of p homeomorphic to the semicircle H, where p is sent to the origin. Any closed path in $H \setminus \{(0, 0)\}$ can be shrunk to a point, the deformation being in $H \setminus \{(0, 0)\}$, so any closed path in $N \setminus \{p\}$ can be shrunk to a point, the deformation being in $N \setminus \{p\}$.

If the point p of a surface with boundary is *not* an edge point then there must be a homeomorphism from some neighbourhood M of p to the open unit disc D in which p is sent to the origin, and we call p an *interior point*. Every neighbourhood N of p contained in M is sent to a neighbourhood N_1 of the origin, which therefore contains a closed path γ whose winding number about the origin is 1. The path γ cannot be shrunk to a point by any deformation in $N_1 \setminus \{(0,0)\}$, so the neighbourhood N contains a closed path that cannot be shrunk to a point by any deformation in $N \setminus \{p\}$. Hence a point of a surface is either an edge point or else an interior point.

Note that a homeomorphism sends edge points to edge points and interior points to interior points.

Example 6.15

A closed Möbius band, that is, with its edge, has a path-connected set of edge points, whereas a closed cylinder does not, so a closed Möbius band is not homeomorphic to a closed cylinder.

A
Continuity

Introduction

In this book we have been studying those topological properties which can be easily visualised. In order to do this we have been drawing pictures on paper. Paper is the wrong medium, however, because it is not elastic, whereas topological spaces are always *extremely* elastic. An elastic space cannot have a notion of distance, because any proposed distance between two points in the space could immediately be altered by stretching the space. Homeomorphisms have been used to define this stretching, and in this appendix we review the definition of continuity on which the idea of homeomorphism depends.

The function $g : \mathbb{R} \to \mathbb{R}$ given by

$$g(x) = \begin{cases} x & x \leq 1 \\ 2x & x > 1 \end{cases}$$

tears the real line, and is *discontinuous* at $x = 1$. Continuous functions can stretch or shrink but not tear the space on which they are defined, which means that their graphs cannot *jump* as in Figure A.1. The function g is continuous for all values of x other than $x = 1$, and it stretches the interval $\{x : 2 < x < 3\}$ into the interval $\{y : 4 < y < 6\}$.

It is not always easy to draw the graph of a function, and so how do we recognise continuous functions without using a graph? We use the idea of a *neighbourhood* of a point. We tend to think of a neighbourhood of a point **p** in

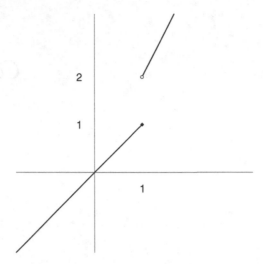

Figure A.1

$\mathbb{R}^n$ as a set of the form

$$\{\mathbf{x} \in \mathbb{R}^n : ||\mathbf{x} - \mathbf{p}|| < r\}$$

for some $r > 0$: when $n = 1$ these sets are the open intervals, while for $n = 2$ or $n = 3$ they are the interiors of discs or spheres respectively.

Our rather more general definition of a neighbourhood is that, given a subset S of $\mathbb{R}^n$ and a point $\mathbf{p}$ in S, a subset N of S is a *neighbourhood* of $\mathbf{p}$ if N *contains* a set

$$\{\mathbf{x} \in S : ||\mathbf{x} - \mathbf{p}|| < r\}$$

for some $r > 0$, that is, N contains all points of S sufficiently close to $\mathbf{p}$.

Returning to our function g, we consider a neighbourhood N of $g(1)$, namely

$$N = \{y \in \mathbb{R} : |y - g(1)| < 1/2\}.$$

Consider the set M of all points on the x-axis which map to N, which we will call the *pre-image* of N :

$$M = \{x \in \mathbb{R} : f(x) \in N\} = \{x \in \mathbb{R} : 1/2 < x \le 1\}.$$

This set M is *not* a neighbourhood of 1 as M contains no points greater than 1.

The tear at $\mathbf{p} = 1$ is reflected in the fact that $g(1)$ has a neighbourhood N whose pre-image is not a neighbourhood of 1. At any point $\mathbf{p}$ other than 1 every neighbourhood of $g(\mathbf{p})$ has a pre-image that *is* a neighbourhood of $\mathbf{p}$. For example, the pre-image of the neighbourhood

$$\{y \in \mathbb{R} : |y - g(5/4)| < 1\}$$

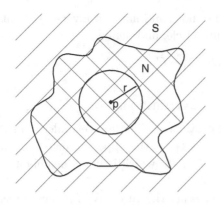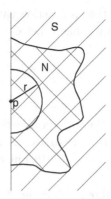

Figure A.2

of $g(5/4)$ is $]1, 7/4[$, which contains

$$\{x \in \mathbb{R} : |x - 5/4| < 1/4\}.$$

Definition A.1

A function $f : S \to T$ between Euclidean sets is *continuous at the point* $\mathbf{p}$ in S if the pre-image of N is a neighbourhood of $\mathbf{p}$ whenever N is a neighbourhood of $f(\mathbf{p})$. The mapping f is *continuous on* S if it is continuous at each point $\mathbf{p}$ in S.

It may seem strange that this characterisation of continuity goes "against" the direction of the function: we started with the neighbourhood N of a point $f(\mathbf{p})$ and considered its pre-image M. What happens if we do things the other way around, and *start* with a neighbourhood M of $\mathbf{p}$?

Consider the continuous function $h : \mathbb{R} \to \mathbb{R}$ given by

$$f(x) = x^2 + 1.$$

The neighbourhood $\{x \in \mathbb{R} : |x| < 1\}$ of the point $x = 0$ has as image the set

$$\{y \in \mathbb{R} : 1 \le y < 2\},$$

which is not a neighbourhood of 1. So we should not expect continuous images of neighbourhoods to be neighbourhoods.

However, for a homeomorphism f we insist that f has an inverse and that both f and f^{-1} are continuous, so for such an f neighbourhoods are mapped to neighbourhoods in either direction. (The function h above has no inverse.)

Note that a function can be continuous and have an inverse without being a homeomorphism. Let C be the unit circle, and define

$$f : [0, 1[\to C$$

by

$$f(x) = (\cos 2\pi x, \sin 2\pi x).$$

The function f is a bijection and continuous. The image under f of $[0, \frac{1}{2}[$—which is the pre-image under f^{-1} of $[0, \frac{1}{2}[$—is the upper semicircle of the unit circle, including $(1, 0)$ but excluding $(-1, 0)$, which is not a neighbourhood of $(1, 0)$.

The above definition of continuity is not the only way of making precise the "no tearing the space, no jumps in the graph" ideas above: there are also well-known, and completely equivalent, "$\epsilon - \delta$" or "limits of sequences" definitions. (See [3] and [2], for example.)

We do not repeat here the development of continuity that shows that basic functions such as $x^2, \cos x, e^x$ and so on are continuous, and that functions suitably built from these, including all functions used in Chapter 1, are continuous.

All topological spaces, whether subsets of $\mathbb{R}^n$ or not, are specified by the allocation of neighbourhoods to each point, so that our definition of continuity given above still makes sense, and indeed remains our definition of continuity in this more general context.

Compactness

We give here the proof, essential for consistency of our discussion of good surfaces, that a subset of a Euclidean space homeomorphic to a closed and bounded Euclidean set is itself closed and bounded. We do this by defining a topological property, called *compactness,* and show that a Euclidean set is compact if and only if it is closed and bounded.

Definition A.2

A topological space X is *compact* if, whenever X is covered by, that is the union of, a collection C of open subsets of X, then X is covered by a *finite* subcollection of the open sets in C.

Theorem A.1

The continuous image of a compact space is compact.

Proof

Let $f : X \to Y$ be continuous and onto, where X is compact. The pre-image under f of an open set in Y is open in X, so a cover C of Y gives rise to a cover D in X. A finite subcollection F of D covers X and the images of the open sets in F cover Y. This completes the proof.

It follows at once that compactness is a topological property.

Theorem A.2

A Euclidean set S is compact if and only if it is closed and bounded.

Proof

The following argument, given for a plane set S, needs little change to apply to any Euclidean set. Suppose that S is not closed, so that there is a point x not in S, every neighbourhood of x meeting S. Consider the set of open discs, centre s, radius $\frac{1}{2}\|s - x\|$, for each s in S. These discs cover S. Given any finite set F of these discs we can find a neighbourhood N of x not meeting any disc in F. But N contains a point of S, so F does not cover S and S is not compact.

Suppose that S is not bounded. Consider the set of open discs, centre s radius 1, for each s in S. These discs cover S. The union of any finite set of such discs is bounded and so cannot cover S. Thus S is not compact. We have shown that a compact Euclidean set is closed and bounded.

Now suppose that S is a subset of the plane that is closed and bounded, but not compact, and that S is covered by the collection C of open sets. Then S is covered by no finite subcollection of C. Now S is contained in some square Q_1. Divide Q_1 into four equal squares. At least one of these four cannot be covered by any finite subcollection of C: call one such square Q_2. Repeat the process of quadrisection to construct a sequence $Q_1, Q_2, Q_3, \ldots$ of squares. Let the bottom left corner of Q_n be (x_n, y_n). Then (x_n) and (y_n) are increasing sequences bounded above, and so converge, to (x, y) say. (This is actually the *monotone convergence theorem*: see [3], page 46.) As there are points of S in every Q_n, and S is closed, (x, y) must be in S, and is in some open set O of C. Some Q_n is covered by the single open set O, a contradiction. So S is compact. This completes the proof.

The result, promised in our discussion of putting circles in the plane, that a continuous mapping f of the whole plane to itself sends a bounded set B to a bounded set, follows from the previous theorems. Now B is contained in some closed disc D, which is closed and bounded, and so compact. From Theorem A.1, the image $f(D)$ is compact, and therefore bounded. Hence $f(B)$ is bounded.

Angle Functions

We give here a brief proof that there is a *continuous* function θ satisfying the equation

$$\gamma(t) = ||\gamma(t)||(\cos\theta(t), \sin\theta(t))$$

at the beginning of Chapter 6. Let M be the least upper bound of all t_0 where such a continuous function can be defined on $[a, t_0]$ and suppose that $a < M < b$. By using either

$$\tan^{-1}\frac{x_2(t)}{x_1(t)} \quad \text{or} \quad \cot^{-1}\frac{x_1(t)}{x_2(t)}$$

we can define a continuous function ϕ on $]M - \epsilon, M + \epsilon[$ for some $\epsilon > 0$ where $\gamma(t) = (x_1(t), x_2(t))$ and ϕ agrees with θ. Extending an angle function θ on $[a, M - \epsilon/2]$ to $[a, M + \epsilon]$ using ϕ contradicts the least upper bound property of M. So $M = b$.

Identification Spaces

We now justify the claim of continuity made in Example 3.6 and make further comment on the torus as an identification space.

It is convenient to study the simpler example of the circle before going on to the torus. Let $\alpha : \mathbb{R} \to S^1$ be the mapping given by

$$\alpha(u) = (\cos 2\pi u, \sin 2\pi u)$$

and suppose that f is a continuous mapping from the real line to a space Y such that $f(u) = f(u_1)$ whenever $u - u_1$ is an integer. Then we may define $\hat{f} : S^1 \to Y$ by $\hat{f}(x, y) = f(u)$ for any u such that $\alpha(u) = (x, y)$. We prove that $\hat{f}$ is continuous at every point $\alpha(u_1)$ in S^1. Put $\alpha(u_1) = (x_1, y_1)$. In the case that u_1 is in an interval $I =]n, n+1/2[$ for some integer n, then the restriction α_1 of α to I is a homeomorphism from I to a semicircle as the inverse of α_1 is the continuous mapping

$$(x, y) \to \frac{\cos^{-1} x}{2\pi}$$

where $\cos^{-1} x_1$ is taken to be $2\pi u_1$. As $\hat{f}(x,y) = f(\alpha_1^{-1}(x,y))$ for all (x,y) in a neighbourhood of (x_1, y_1) it follows that $\hat{f}$ is continuous at (x_1, y_1).

Similarly, if u_1 is in an interval of the form

$$]n + 1/4, n + 3/4[, \quad]n + 1/2, n[, \quad]n - 1/4, n + 1/4[$$

we have respective continuous inverses

$$(x,y) \to \frac{\sin^{-1} y}{2\pi}, \quad (x,y) \to \frac{\cos^{-1} x}{2\pi}, \quad (x,y) \to \frac{\sin^{-1} y}{2\pi}.$$

Hence $\hat{f}$ is continuous.

We now indicate how to extend this result to the torus. Put

$$\beta(u,v) = (\cos 2\pi u, \sin 2\pi u, \cos 2\pi v, \sin 2\pi v)$$

and suppose that g is a continuous mapping from the plane to a space Y such that $g(u,v) = g(u_1, v_1)$ whenever $u - u_1$ and $v - v_1$ are integers, so we may define $\hat{g}(x,y,z,w)$ to be $g(u,v)$ for any (u,v) such that $\beta(u,v) = (x,y,z,w)$.

In the case that (u_1, v_1) is in a square

$$Q = \{(u,v) : n < u < n + 1/2, m < v < m + 1/2\},$$

where m, n are integers, the restriction β_1 of β to Q is a homeomorphism from Q to its image as β_1 has a continuous inverse of the form

$$(x,y,z,w) \to \left(\frac{\cos^{-1} x}{2\pi}, \frac{\cos^{-1} z}{2\pi} \right).$$

As $\hat{g}(x,y,z,w) = g(\beta_1^{-1}(x,y,z,w))$ for all (x,y,z,w) in a neighbourhood of $\beta(u_1, v_1)$ it follows that $\hat{g}$ is continuous at $\beta(u_1, v_1)$.

In all cases we consider a square Q whose sides are one of the four intervals given above, and the continuity of $\hat{g}$ at $\beta(u_1, v_1)$ follows as before after appropriate changes to the formula for the inverse of β_1.

To apply the above result to prove the continuity of the mapping g^* of Example 3.6, note that g^* is $\tau \hat{g} \sigma^{-1}$, where σ is stereographic projection from the torus T^2 to the torus T_S.

An argument similar to the above shows that a continuous mapping $h : S^2 \to Y$ in which opposite points of the sphere are mapped to the same point of Y gives rise to a *continuous* mapping from the real projective plane to Y. Here continuity can be shown by considering the restrictions of h to each of the six open hemispheres formed by cutting the sphere by the three coordinate planes: details are omitted.

We now work towards a proof that the torus is homeomorphic to the plane with identification. We first prove the general result that, if $\hat{S}$ is any identification space and $f : S \to Y$ is continuous and constant on each $\hat{s}$, then the mapping $\hat{f} : \hat{S} \to Y$ given by $\hat{f}(\hat{s}) = f(s)$ for any s in $\hat{s}$ is continuous.

Take any $\hat{a}$ in $\hat{S}$ and any neighbourhood N of $\hat{f}(\hat{a})$. As f is continuous, $f^{-1}(N)$ is a neighbourhood of any point s in S such that $f(s) = \hat{f}(\hat{a})$ and is therefore a neighbourhood of all points in $\hat{a}$. From the definition of neighbourhoods in an identification space, because $f^{-1}(N)$ is the union of all the classes in $\hat{f}^{-1}(N)$, it follows that $\hat{f}^{-1}(N)$ is a neighbourhood of $\hat{a}$, and $\hat{f}$ is continuous.

In particular if $\hat{\mathbb{R}}$ is the real line with u related to u_1 if and only if $u - u_1$ is an integer, $Y = S^1$ and

$$\alpha(u) = (\cos 2\pi u, \sin 2\pi u),$$

then $\hat{\alpha} : \hat{\mathbb{R}} \to S^1$ is continuous. Further, if $\rho(u)$ is the class containing u, then from the earlier discussion $\hat{\rho} : S^1 \to \hat{\mathbb{R}}$ is continuous. But $\hat{\alpha}, \hat{\rho}$ are inverse, so S^1 is homeomorphic to $\hat{\mathbb{R}}$.

Similarly, if we consider the plane with (u, v) related to (u_1, v_1) if and only if $u - u_1$ and $v - v_1$ are integers, and $\rho(u, v)$ is in the class containing (u, v), then $\hat{\beta}, \hat{\rho}$ are continuous and inverse, so the torus is homeomorphic to the identified plane.

B
Knots

In Chapter 3 we explored non-equivalent ways of putting circles into the plane or the sphere. Now we ask: how many different ways are there of putting a circle in *space*? This is a much harder question, and in this appendix we will not attempt to prove very much. Instead we content ourselves with introducing the *Jones polynomial*—a knot invariant discovered in 1984 which has stimulated a rapid growth in the subject.

An example of a circle in space is the *trefoil* shown in Figure 3.10 as a $(2,3)$-circle. Intrinsically it is a circle, but it has been put in space in a knotted way. If instead we take a *geometrical* circle in space, such as

$$\{(x, y, z) : x^2 + y^2 = 1, z = 0\},$$

we have the *unknot U*. In general, a *knot* is the image set of a continuous one-one mapping

$$\gamma : S^1 \to \mathbb{R}^3.$$

A rubber band may be thought of as a physical embodiment of the unknot, and it is still the unknot no matter how we stretch or bend it. This type of deformation is called *ambient isotopy*: it is a version of the deformation we used in Chapter 6, modified to prevent knots from being unknotted by pulling tightly. If two knots are ambient isotopic then we regard them as being the same as each other. (Also, if they are ambient isotopic then they must be equivalent in space, but the converse is not true: an example is a trefoil knot and its reflection.)

The essential task is this: given two knots decide, and then show, whether or not they are ambient isotopic.

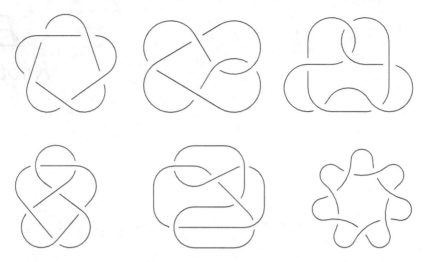

Figure B.1

Our first step is to agree on a convention for drawing knots. We imagine projecting our knot into the plane of the paper, and we indicate an undercrossing by putting a small break in the arc, just as in Figure 3.10. We have to make sure that there are no "triple" crossings, where two arcs cross a third one at the same point. Some examples are given in Figure B.1. Now our problem is this: given two such diagrams, are they ambient isotopic or not?

Suppose that we believe that two diagrams *do* represent ambient isotopic knots, and we want to prove it. Writing down a formula for an ambient isotopy would be extremely difficult, even for fairly simple knots. Fortunately, a very powerful theorem dramatically simplifies this problem.

Theorem B.1 (Reidemeister)

Two knot diagrams D_1 and D_2 represent ambient isotopic knots if and only if D_1 can be changed into D_2 by a finite sequence of the moves RI, RII, and RIII illustrated in Figure B.2.

We omit the proof, and instead suggest the following as an exercise: show that the two knots in Figure B.3 are ambient isotopic.

Note that RI and RII can be used to reduce the number of crossings in a diagram. The smallest number of crossings in all diagrams of a given knot is called the *minimum crossing number* of that knot. The examples in Figure B.1 all have crossing number 7 or less.

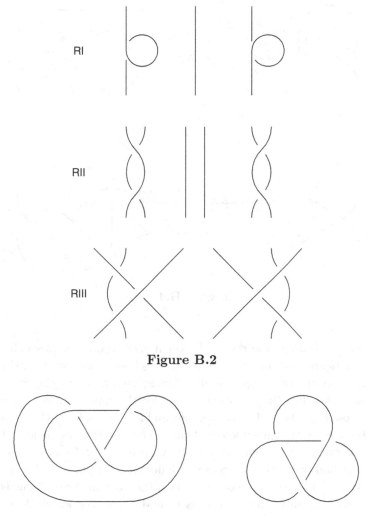

Figure B.2

Figure B.3

There is a fairly obvious way of *composing* two knots, which we illustrate in Figure B.4. We use the notation $K_1 \sharp K_2$ for the composition of K_1 and K_2.

Any knot K composed with the unknot U is just K again, of course. In this sense the unknot is acting like the number 1 under multiplication. It is not possible to undo a knot $K(\neq U)$ by composing with another knot: there is never an "inverse" K^{-1} such that $K \sharp K^{-1} = U$. So knots are either composite or prime: a *prime* knot K is not ambient isotopic to any composition $K_1 \sharp K_2$ unless one of K_1 or K_2 is the unknot. In fact all the knots in Figure B.1 are prime.

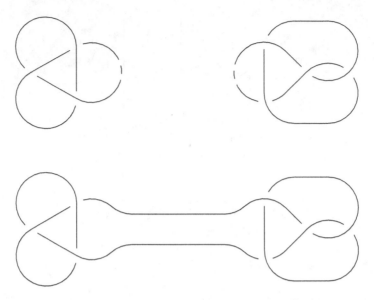

Figure B.4

In the classification of surfaces (Theorem 5.2) we obtained two elegant sequences: knots are much more complicated. It is known, for example, that there
are 1,701,936 prime knots up to and including minimum crossing number 16.
But in any work like this, or even to answer the much simpler question of
how we know that the trefoil knot is not ambient isotopic to the unknot, we
need to be able to distinguish knots. That is, suppose that we believe that two
knots are *not* ambient isotopic: how would we prove this? Just as in our earlier
chapters, we find properties of knots which do not alter under ambient isotopy.
Then, if two knots have different properties, they cannot be ambient isotopic.
By Theorem C.1, all we have to do is to find properties which do not alter
under the Reidemeister moves.

There are many such properties. Some are numbers (such as the minimum
crossing number), while others are polynomials or groups. We restrict ourselves
here to a brief account of the Jones polynomial, which is actually an invariant
of *oriented* knots. To orient a knot simply put an arrow on it, thus determining
a direction around the S^1. For any knot K the Jones polynomial $p(K)$ is a
polynomial in x such that if L is ambient isotopic to K then $p(K) = p(L)$.
In other words, if $p(L) \neq p(M)$ then the knots L and M cannot be ambient
isotopic. We will have to allow these polynomials to be *Laurent* polynomials,
that is, they are allowed to have negative powers of x as well.

The Jones polynomial is defined by three conditions: its value on the unknot,
its value on the disjoint union of a knot with the unknot, and a "skein relation".

The first is easy—we choose its value on the unknot to be 1:

$$p(U) = 1.$$

For the second, let $U \sqcup K$ denote the disjoint union of U and K. This means that the knots U and K are placed in space in such a way that they can be separated by a flat plane. In fact this is not a knot, it is an example of an *oriented link*, because it has more than one component: here two circles have been put into space, not just one. Our entire discussion of the Jones polynomial could have been in terms of links, with knots as the special case of links with just one component. In particular, the Jones polynomial is an invariant of oriented links. We choose:

$$p(U \sqcup K) = -(x + x^{-1})p(K).$$

Finally, we need the *skein relation*. Suppose K^+, K^-, and K^0 are three knots with identical diagrams except at one crossing, where they are as drawn in Figure B.5. Here K^0 is the only way of removing the crossing in K^+ consistent with the arrows. Then

$$x^{-2}p(K^+) - x^2p(K^-) + (x^{-1} - x)p(K^0) = 0.$$

We should show that p does not change under the (oriented) Reidemeister moves. In fact move RI is not too difficult, and we suggest it as an exercise. However, the others are hard—Jones had a completely different way of defining this polynomial and knew in advance that it had to be a knot invariant. A fairly elementary way of demonstrating invariance under the Reidemeister moves is to derive the Jones polynomial from the "Kauffman" polynomial, as in [1].

Let us instead calculate the Jones polynomial of the trefoil knot. The original knot is denoted by T^+ in Figure B.6, which also shows the corresponding T^- and T^0.

Using RII and then RI reduces T^- to U, and we know $p(U)$. So if we can calculate $p(T^0)$ we will be able to use the skein relation

$$x^{-2}p(T^+) - x^2p(T^-) + (x^{-1} - x)p(T^0) = 0$$

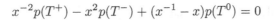

K^+ K^- 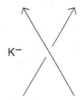 K^0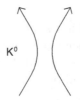

Figure B.5

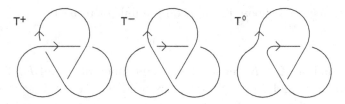

Figure B.6

to obtain $p(T^+)$. Let us relabel T^0 as L^+, shown in Figure B.7 with the corresponding L^- and L^0. Using RI reduces L^0 to U, and using RII reduces L^- to

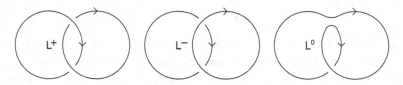

Figure B.7

$U \sqcup U$, which from the second and first defining conditions has polynomial

$$p(U \sqcup U) = -(x + x^{-1})p(U) = -(x + x^{-1}).$$

Now the skein relation for L^+, L^- and L^0 is

$$x^{-2}p(L^+) - x^2 p(L^-) + (x^{-1} - x)p(L^0) = 0,$$

from which we can deduce (after a little algebra, left to the reader) that

$$p(L^+) = -x^5 - x.$$

But this is also $p(T^0)$, and so we return to the skein relation for T^+, T^- and T^0, which becomes

$$x^{-2}p(T^+) - x^2 + (x^{-1} - x)(-x^5 - x) = 0.$$

Again, a little algebra reveals that

$$p(T^+) = x^2 + x^6 - x^8.$$

We noted in Chapter 3 that the $(2,3)$-circle on the torus was a trefoil knot. Similarly, whenever m and n have no common factors we can draw an (m, n)-circle on the torus and then remove the torus to obtain another knot. Its Jones polynomial will be

$$x^{(m-1)(n-1)}(1 - x^{2(m+1)} - x^{2(n+1)} + x^{2(m+n)})/(1 - x^4),$$

although this is not easy to show.

Good though it is, the Jones polynomial does not always succeed in distinguishing knots: the pair shown in Figure B.8 are known not to be ambient isotopic, but they have the same Jones polynomial. So in general we may have to cast around for other invariants, just as we did with the cut-points, cut-pairs, and so on in Chapter 2.

Let us end with one of the many fascinating open questions here: is there a knot K (other than the unknot) which has $p(K) = 1$?

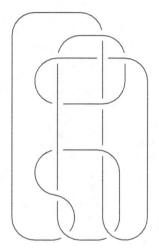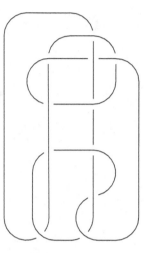

Figure B.8

C
History

This appendix is very far from being a comprehensive account of the early history of topology. For that we recommend [7], on which we depended in writing this. (In particular, our quotations are free translations from [7].) We concentrate on the early history because that is when the ideas in this book were discovered; much of the rapid growth in the subject since then has been in *algebraic* topology, in which one uses algebraic structures—such as groups—as topological invariants.

The first time Euler mentioned his theorem that

$$F - E + V = 2$$

was in a letter to Goldbach on the 14th of November 1750. His proof was the first of many erroneous ones: in fact it was not until 1847 that von Staudt gave the graph-theoretic argument we use. Euler's idea was to show that given a polyhedron, certain pieces could be cut off without changing the value of $F - E + V$, in such a way that one eventually arrived at a tetrahedron, for which $F - E + V = 2$. The proof fails, because it is possible to arrive at two polyhedra joined only along an edge, for example (see [5]).

A completely different proof was given by Legendre in 1794. Given a convex polyhedron choose any point inside it and a sphere centred on that point. Then project the polyhedron onto the sphere. Each face of the polyhedron has now become a spherical polygon, and it only remains to set the area of the whole sphere equal to the sum of the areas of these spherical polygons. When one does this (and divides by 4π) Euler's theorem appears. In fact, this method can also be made to work for a non-convex polyhedron, as was explicitly noted in 1809 by Poinsot, but it is not a *topological* proof.

In 1811 Cauchy tried to construct a proof by removing one face from the polyhedron and then regarding the remainder as a planar network. It is as though, for a spherical polyhedron, he used stereographic projection. However, he was not aware that he had restricted himself to spherical polyhedra.

The first person to make it perfectly clear that there are different types of polyhedra, each type having a different value for $F - E + V$, was Lhuilier. He argued that if we are given two polyhedra each of which satisfies $F - E + V = 2$, and such that a face on one is equal to a face on the other, then the polyhedron obtained by gluing these two faces together also satisfies $F - E + V = 2$. Given a polyhedron, it therefore suffices to choose any point inside it and from that point divide the polyhedron into pyramids. He accepted, however, that this would fail if the starting polyhedron had a cavity, and he went further: if this cavity is itself a polyhedron with $F - E + V = 2$ then the original polyhedron will satisfy $F - E + V = 4$. Lhuilier also considered the (more familiar) "pathology" of the polyhedron having a hole (in the sense of being homeomorphic to a torus), or even more than one such hole. He showed that these polyhedra satisfy $F - E + V = -2(n - 1)$, where n is the number of holes, thus introducing the concept of the genus for the first time. All this was published in 1813.

von Staudt's proof of 1847 gave correct conditions for a polyhedron to satisfy $F - E + V = 2$. Our proof is a reformulation of von Staudt's. In 1851 Schläfli generalised the theorem to n-dimensional polyhedra.

It is significant that Gauss was aware of the crucial importance of topology, and yet was unable to do more than make some characteristically original isolated discoveries. In 1799 Gauss's proof of the fundamental theorem of algebra used a topological argument, for example. Also, although Gauss made enormous contributions to cartography, he was perfectly clear that the class of mappings used there was part of a much larger class:

> Besides, these are only special cases of the more general representation of one surface on another which to each point of one relates a point of the other, and this in a continuous way.

However, perhaps his most lasting contribution to topology was the strong encouragement he gave to Möbius and Listing.

His student Listing coined the word "topology", in a letter of 1836, and his *Vorstudien zur Topologie* (1847) was the first book to have the word topology in the title. His *Census räumlicher Complexe oder Verallgemeinerung des Euler'schen Satzes von den Polyedern* (1861) recognized for the first time that Euler's theorem was a part of topology. In it he showed how to define an important generalisation of a polyhedron called a "complex", although it is arguably a shame that Listing chose to highlight this in his title: in his notes for this work we find various alternatives, such as *Topological generalisation of Euler's Theorem on Polyhedra.*

Riemann's inaugural dissertation of 1851 *Foundations for a general theory of functions of one complex variable* gave mathematicians a powerful new incentive to study topology. It was no longer merely a question of properly understanding Euler's theorem: Riemann showed that the theory of complex functions and their integrals depended upon some essentially topological concepts. He called a region in a surface "simply-connected" if all closed paths within it can be deformed to a point. He then showed how to divide a surface into simply-connected pieces by drawing extra "transverse sections" between their boundaries. Finally, he defined the "order of connection" p of a surface in terms of the numbers of simply-connected pieces and transverse sections, showing that for a given surface p does not depend on how it is divided up. These brilliant early ideas were the basis for later work by Betti and Poincaré which led to one of the key tools of modern topology—homology theory.

In 1861 Möbius defined what we now call a homeomorphism between two spaces (he was thinking of surfaces) as follows:

> ... each point of one corresponds to a point of the other, in such a way that two infinitely neighbouring points always correspond to two infinitely neighbouring points ...

He went on to say:

> If for example we imagine a perfectly flexible and elastic spherical surface, all the possible forms into which we can put it by bending and stretching (without tearing), will be mutually homeomorphic. On the other hand, a spherical surface and the surface of a torus are not homeomorphic.

This second paragraph certainly helps us to see some homeomorphisms, but in fact a homeomorphism between two surfaces does *not* require that one can be deformed into the other. Indeed Möbius was perfectly well aware of this, as his definition shows, and this distinction was a very important step in the evolution of topology.

These ideas were in Möbius's submission to the Paris Academy of Sciences for the *Grand Prix de Mathématiques* of that year, which he did not win, in spite of the fact that this submission also included the theorem classifying all closed orientable surfaces. He put these surfaces into "types" as follows:

> ... a closed surface of type n ... can be divided into two primitive forms by n closed curves ...

On a closed surface of type n one can construct $n-1$ closed curves which do not divide it.

(By "primitive form" Möbius meant a sphere with a number of discs removed.) His theorem, the *fundamental* theorem of this subject, then states that two surfaces of the same type are homeomorphic, while two surfaces of different types cannot be homeomorphic.

He went on to discuss Euler's theorem in terms of this number n, and showed that

$$V + F = E - 2(n-2).$$

We can see that n is one more than our genus k.

However, Möbius is best known for his band. It is particularly ironic therefore that it was Gauss who described the "Möbius band" to Möbius, sometime before 1858. Its name is reasonable, though: Möbius systematically studied triangulations of one-sided surfaces like his band.

Jordan gave a very similar definition of homeomorphism in 1866, independently of Möbius. This is an example of a common phenomenon, whereby a new idea appears to be "in the air", and two or three mathematicians grasp it nearly simultaneously. Jordan went on to give a detailed study of the classes of closed curves which can be drawn on a surface, where two curves are in the same class when one can be deformed into the other. He very nearly arrived at what we now know to be the fundamental group.

In Klein's "Erlanger Programme" of 1872 he argued that topology was the study of those properties of spaces which are invariant under (the group of) homeomorphisms. This was part of a much more general synthesis of many branches of geometry which has had a profound effect on the subject. Indeed, in 1873 Clifford wrote:

> The setting up of a correspondence between two sets, and the study of those properties which are conserved by this correspondence, can be regarded as the central idea of modern mathematics ...

Klein acknowledged the importance of the work of Möbius, especially in making clear the various different sorts of mappings between surfaces, of which homeomorphism is one. (Another key ingredient was the rediscovery of Galois' 1846 work on groups.)

In around 1873 Klein and Schläfli studied the real projective plane from the topological point of view, applying Riemann's order of connection idea to it. However, it was not until 1876 that Klein gave the following definition of a one-sided surface:

> One draws on a part of the surface a closed curve, on which an orientation is chosen; the surface is then one-sided if and only if it is possible to move this curve around in the surface in such a way that its orientation is reversed when it returns to its original position.

Klein gave a beautifully clear account of the ideas of Riemann and Jordan in a course given in Leipzig in 1880–1881. In them we also find an early recognition of the importance of the discovery (in 1877) by Cantor of some highly counter-intuitive "space-filling curves" acting as a warning that the topological definition of dimension was not at all clear. (Jordan's proof, in 1887, that a continuous closed curve in the plane divides it into two regions was a response to the prevailing culture of caution which followed Cantor's discoveries.) Finally we should note that it was in these Leipzig lectures that Klein first described his bottle.

Even more than his explicit contributions to topology, Klein's strong encouragement for the rapid development of the theory of complex functions and Riemann surfaces had the effect of encouraging the equally rapid development of topology. However, this story would take us well beyond the scope of the book, so instead we end this appendix with a few remarks on the history of knot theory.

Gauss's papers include a collection of drawings of knots (1794), and some notes (of 1833) in which he gives an extraordinary integral formula for the linking number of two space curves (see the excellent article by M. Epple, *Geometric Aspects in the Development of Knot Theory*, pages 301–357 of [6]). These discoveries were partly motivated by his studies of the orbits of asteroids and by his work on the differential geometry of surfaces, and, as Epple argues, the high expectations which Gauss had for the future subject of topology must have come from his continually finding topological objects in a variety of disciplines within pure mathematics or the physical sciences.

The next significant step forward was made by Tait, Kirkman, and Little, who between 1880 and 1900 compiled some extraordinary tables of knots. There were no knot invariants yet, so these tables contain some duplicates: what is remarkable is that there are so few. These tables were part of an attempt to use the dynamics of a continuous medium to explain the existence of atoms: this tension between the continuous and the discrete is still evident in modern mathematical physics.

The classification problem for knots was first explicitly stated by Maxwell, who also gave a clear account of the physical content of Gauss's linking integral in his treatise on electromagnetism of 1873.

Reidemeister's theorem was proved in 1926 (but his moves had in fact already been written down by Maxwell) and then in 1928 Alexander was able to use it to establish the first polynomial knot invariant, which he tested on Tait's tables. He also noted that his polynomial obeyed what we now call a skein relation, but it was Conway who in 1967 first saw the possibility of *defining* a polynomial knot invariant from its skein relation. Following the discovery of the Jones polynomial in 1984, skein relations have been used as a very effective way of exploring knot invariants.

1.1 Turn the interval over, stretch by a factor of four, and slide along so 0 is mapped to 7, giving the formula

$$x \to 7 - 4x.$$

The inverse is

$$x \to \frac{1}{4}(7 - x).$$

1.2 Combining Examples 1.5 and 1.6 gives the homeomorphism

$$x \to \frac{2x - 1}{1 - |2x - 1|}.$$

Another is

$$x \to \log\left(\frac{1}{x} - 1\right).$$

1.3 A suitable homeomorphism from the first to the second rectangle is

$$(x, y) \to \left(2y, \frac{x}{2}\right),$$

the inverse having the same formula.

1.4 Map each point radially outwards so that its distance from the origin is what it was before, plus one,

$$\mathbf{x} \to (||\mathbf{x}|| + 1)\mathbf{x}/||\mathbf{x}||,$$

the inverse formula being

$$\mathbf{x} \to (||\mathbf{x}|| - 1)\mathbf{x}/||\mathbf{x}||.$$

The same formulae show that the plane with the origin removed is homeomorphic to the plane with the disc D removed.

1.5 Project the cone down to the plane, by

$$(x, y, z) \to (x, y).$$

The inverse mapping is given by

$$(x, y) \to (x, y, \sqrt{x^2 + y^2}).$$

1.6 First shrink the hyperboloid horizontally to an infinite cylinder using

$$(x, y, z) \to \left(\frac{x}{\sqrt{x^2 + y^2}}, \frac{y}{\sqrt{x^2 + y^2}}, z \right),$$

with inverse

$$(x, y, z) \to (x\sqrt{1 + z^2}, y\sqrt{1 + z^2}, z).$$

Now shrink vertically using

$$(x, y, z) \to (x, y, f(z)),$$

where f is some homeomorphism from the real line to $]-1, 1[$, with inverse

$$(x, y, z) \to (x, y, f^{-1}(z)).$$

(The homeomorphism f could be

$$f(x) = \frac{x}{1 + |x|},$$

for example.) This shows that the hyperboloid is homeomorphic to the given cylinder. Deform the cone horizontally, using

$$(x, y, z) \to \left(\frac{x}{\sqrt{x^2 + y^2}}, \frac{y}{\sqrt{x^2 + y^2}}, z \right),$$

to the cylinder $\{(x, y, z) : x^2 + y^2 = 1, z > 0\}$. Now expand vertically to the infinite cylinder above using

$$(x, y, z) \to (x, y, \log z).$$

The inverse mappings are respectively

$$(x, y, z) \to (xz, yz, z)$$

and
$$(x, y, z) \rightarrow (x, y, e^z).$$

Expand the sphere to the given cylinder using

$$(x, y, z) \rightarrow \left(\frac{x}{\sqrt{x^2 + y^2}}, \frac{y}{\sqrt{x^2 + y^2}}, z \right).$$

The inverse mapping is

$$(x, y, z) \rightarrow (x\sqrt{1 - z^2}, y\sqrt{1 - z^2}, z).$$

1.7 As in Example 1.7, taking h to be a homeomorphism from $]0, 1[$ to the real line, a homeomorphism from C to space is given by

$$(x, y, z) \rightarrow (h(x), h(y), h(z)).$$

Example 1.5 shows that B is homeomorphic to space, so C is homeomorphic to B.

1.8 A homeomorphism from the first to the second triangle is

$$(x, y) \rightarrow (x, y - (1 - |x|)),$$

with inverse

$$(x, y) \rightarrow (x, y + (1 - |x|)),$$

and from the second to the third is

$$(x, y) \rightarrow (x, -y),$$

the inverse having the same formula.

1.9 Construct f from the square to the disc by mapping radially and linearly, so that, for $|y| \le |x| \ne 0$ we have

$$f(x, y) = (x, y) \frac{|x|}{\sqrt{x^2 + y^2}},$$

with inverse

$$g(x, y) = (x, y) \frac{\sqrt{x^2 + y^2}}{|x|},$$

and similarly for $|x| \le |y| \ne 0$. Noting that the formulae for f in the two regions agree along the cuts $|y| = |x|$, and putting $f(0, 0) = (0, 0)$, we have a homeomorphism from the square to the disc.

1.10 For each $r > 0$ denote by S_r^2 the sphere of radius r centre the origin, with the north pole removed. Now S_r^2 is mapped homeomorphically to $S^2 \setminus \{(0,0,1)\}$ by

$$(x,y,z) \rightarrow \left(\frac{x}{r}, \frac{y}{r}, \frac{z}{r}\right),$$

and spiked space is the union of all the S_r^2 for $r > 0$.

Using stereographic projection

$$(x,y,z) \rightarrow \left(\frac{x}{1-z}, \frac{y}{1-z}\right)$$

we get a homeomorphism from spiked space to $\{(x,y,z) : z > 0\}$ with formula

$$(x,y,z) \rightarrow \left(\frac{x/r}{1-z/r}, \frac{y/r}{1-z/r}, r\right),$$

where $r = \sqrt{x^2 + y^2 + z^2}$.

But

$$(x,y,z) \rightarrow (x,y,\log z)$$

is a homeomorphism from $\{(x,y,z) : z > 0\}$ to $\mathbb{R}^3$, so spiked space is homeomorphic to the whole of space.

2.1 The twelve sets shown in Figure D.1 have all n-points for $n \geq 3$ marked. Sets having not-cut-points are marked with an N. No two sets have the same number of n-points for all n, so no two are homeomorphic. The convention here is that the end of a line segment without other indication is *missing*. The first set can either be constructed as the union of three identical segments or, for example, as the union of $]0,3[$, $]2,5[$, and $]4,7[$.

Any set satisfying the given conditions is homeomorphic to one of the twelve sets shown. The three sets shown in Figure D.2 are *not* extra solutions. The first is not path-connected, the segment PQ being one of the two components (see Example 2.6). The second set is not joined at R and is homeomorphic to the example having two 4-points. The third set is homeomorphic to the example with two 3-points.

2.2 Reinterpret the segments in Figure D.1 as *having* their end-points: the N now means having infinitely many not-cut-points. In addition there are the sets shown in Figure D.3.

Sets having the same number of n-points for all n have *different* numbers of components of 2-points—of the sets marked N each "arm" gives one component of 2-points.

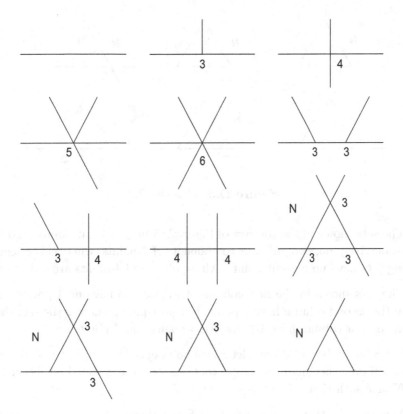

Figure D.1 Exercise 2.1

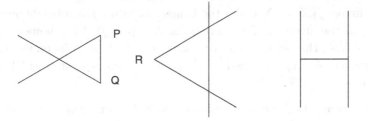

Figure D.2 Not extra solutions to Exercise 2.1

2.3 The sets shown in Figure D.4 have all 3-points and 4-points marked. The two sets with a 4-point but no 3-point have their 2-points indicated by heavier lines. One set has five components of 2-points whereas the other has only four. The four sets with a 3-point but no 4-point have their not-cut-points indicated by heavier lines. The four sets have respectively two, three, four, and five components of not-cut-points.

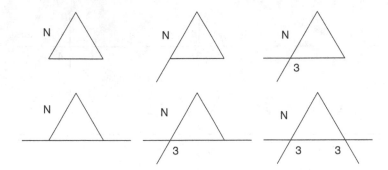

Figure D.3 Exercise 2.2

2.4 The sets shown in the top row of Figure D.5 have not-cut-points, and have respectively no, two and four components of 2-points, whereas the remaining sets have no not-cut-points. All 3-points and 4-points are indicated.

2.5 The sets shown in the first column of Figure D.6 have no 1-points, those in the second column have 1-points but no 1-pairs, whereas the sets shown in the last column have 1-pairs. All 3-points and 4-points are indicated.

2.6 Let S be a subset of $\mathbb{R}^n$ and let x be a point of S. We say that x is an *n-node* of S if every neighbourhood of x contains a path-connected neighbourhood N of x such that x is an n-point of N.

Now suppose that x is an n-node of S and that $f : S \to T$ is a homeomorphism. Let M be a neighbourhood of $f(x)$. The pre-image of M contains a path-connected neighbourhood N such that x is an n-point of N. But the image $f(N)$ of N under the homeomorphism f is a neighbourhood of $f(x)$ as the inverse of f is continuous. Also $p \to f(p)$ is a homeomorphism $N \to f(N)$ that sends x to $f(x)$, so that $f(N)$ is a path-connected neighbourhood of $f(x)$ contained in M and $f(x)$ is an n-point of $f(N)$. So $f(x)$ is an n-node of T.

The eleven sets shown in Figure D.7 have different numbers of 4-nodes or 6-nodes except for a pair of sets having three 4-nodes, only one of which has a 2-point, and a pair of sets having four 4-nodes, only one of which has a 4-pair.

3.1 Figure D.8 shows nine suitable subsets of S with their complements in S shown below. The complements of A, B, C are path-connected and have infinitely many not-cut-points in respectively 4,2,1 components. The complements of D, E, F are path-connected and have respectively 2,1,1 not-cut-points. The closure of F is a circle but the closure of E is not. The complements of G, H, I all have two components, but no two complements

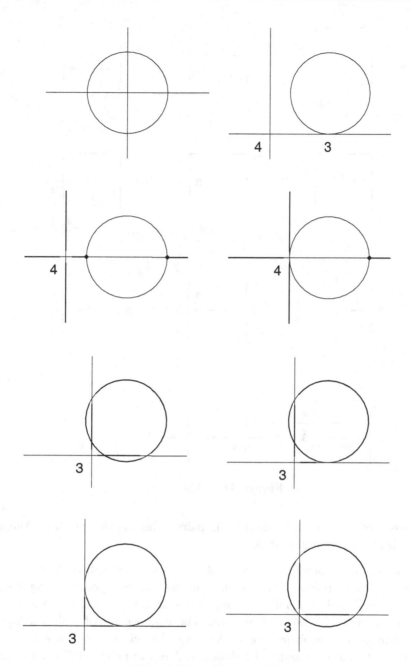

Figure D.4 Exercise 2.3

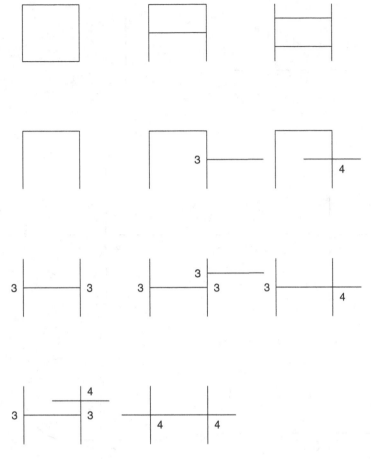

Figure D.5 Exercise 2.4

have components homeomorphic in pairs. It follows that no two of the nine
subsets are equivalent in S.

3.2 Figure D.9 shows fourteen suitable subsets of the sphere. Counting the
number of 2-points of the closure, the number of 3-points of the closure,
and the number of infinite components of not-cut-points of the closure
distinguishes all except five pairs. The complement of A has a 2-point
whereas the complement of H does not. The closure of J has a not-cut-
pair whereas the closure of I does not. Any two points of C can be joined
by a path that does not meet the closure except at its ends, whereas this
is not true of B, and similarly for D, E and F, G.

Note that the sets K, L of Figure D.10 are *not* extra solutions: the sets K
and C are equivalent in the sphere, as are L and F.

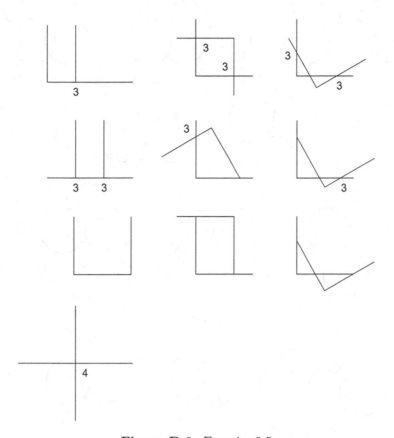

Figure D.6 Exercise 2.5

3.3 Figure D.11 shows eight suitable subsets of the Möbius band. We first count
the number of times the two circles of the figure eight go round the Möbius
band: 0,0 for A and B; 0,1 for C; 0,2 for D and E; 1,1 for F; 1,2 for G;
and 2,2 for H. To distinguish A from B we consider the two components
of the complement that are discs. In the case of A the intersection of the
closures of these two components is a single point whereas in the case of B
it is not. To distinguish D from E we apply the same argument as above to
the two components of the complement that are homeomorphic to a disc
or a cylinder.

3.4 Figure D.12 shows six suitable subsets of C with the closure of each
shown below. The subsets E, F have homeomorphic closures but E has
a path-connected complement whereas F does not. The subsets B, C have
homeomorphic closures but the complement of the closure of B has a

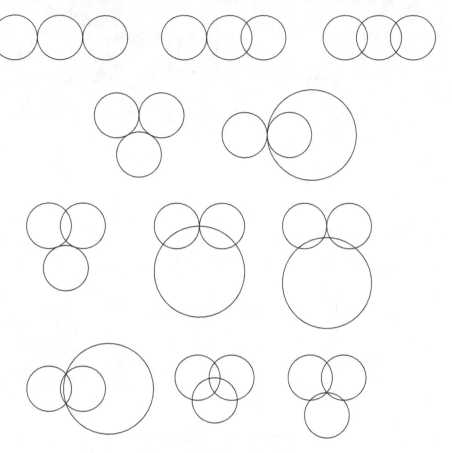

Figure D.7 Exercise 2.6

component homeomorphic to a disc, whereas the complement of the closure of C does not.

4.1 Rest the cube on a horizontal surface, then press flat while simultaneously stretching the top four edges outwards, leaving the bottom edges fixed. We now have a sphere with five holes, or handles.

4.2 See Figure D.13.

5.1 As there are at least five edges at each vertex we have $5V \leq 2E$. It is always true that

$$3F_3 + 4F_4 + 5F_5 + \cdots = 2E$$

and

$$F_3 + F_4 + F_5 + \cdots = F,$$

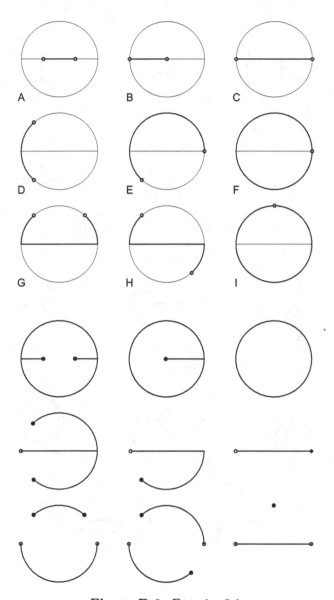

Figure D.8 Exercise 3.1

and, on the sphere,

$$F - E + V = 2.$$

So

$$2 \leq F - E + \frac{2}{5}E = F - \frac{3}{5}E$$

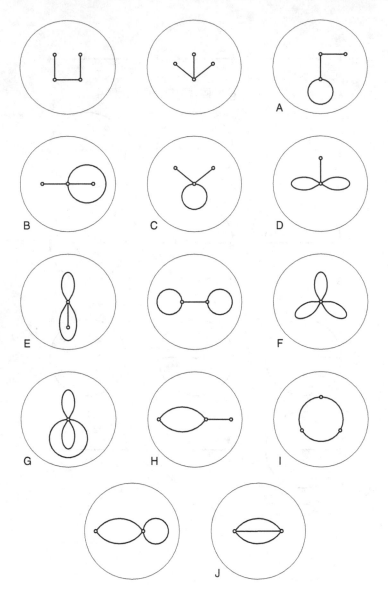

Figure D.9 Exercise 3.2

and

$$10F - 6E \geq 20.$$

Hence

$$10(F_3 + F_4 + F_5 + \dots) - 3(3F_3 + 4F_4 + 5F_5 + \dots) \geq 20$$

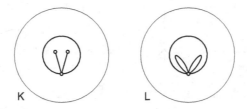

Figure D.10 Not solutions to Exercise 3.2

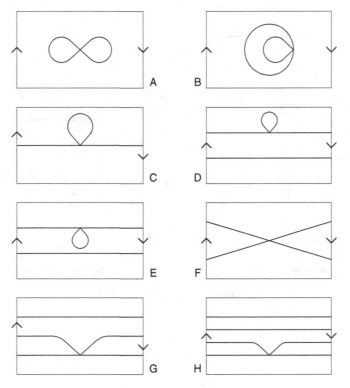

Figure D.11 Exercise 3.3

and so

$$F_3 \geq 20.$$

5.2 We have

$$F_4 + F_6 = F,$$

$$4F_4 + 6F_6 = 2E,$$

$$F - E + V = 2,$$

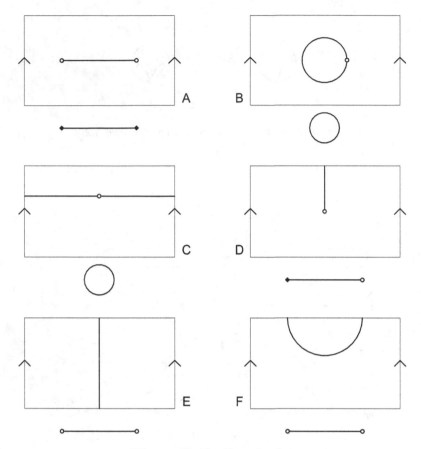

Figure D.12 Exercise 3.4

and

$$3V = 2E.$$

Now

$$2 = F - E + V = F - E + \frac{2}{3}E = F - \frac{1}{3}E$$

so $6F - 2E = 12$ and $6(F_4 + F_6) - (4F_4 + 6F_6) = 12$. Hence $F_4 = 6$.

The first four equations are satisfied by several polyhedra, including the hexagonal prism and cube, but a polyhedron with one square at each vertex satisfies $4F_4 = V$, so that $V = 24$, $2E = 72$. Hence $6F_6 = 72 - 24 = 48$ and $F_6 = 8$.

We construct such a polyhedron by cutting off the corners of an octahedron, giving the truncated octahedron, shown in Figure D.14.

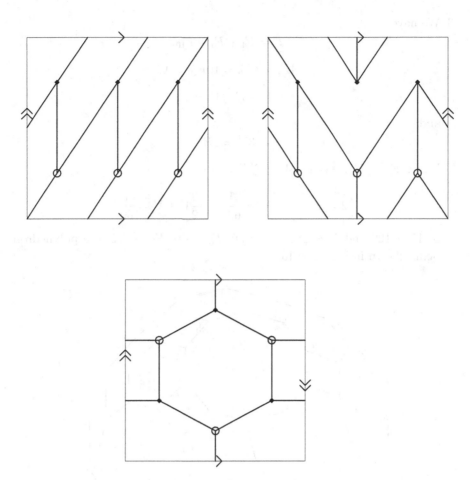

Figure D.13 Exercise 4.2

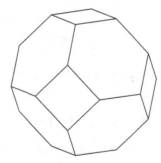

Figure D.14 Truncated octahedron

5.3 We have

$$F = F_4 + F_6 + F_{10},$$

$$4F_4 = 6F_6 = 10F_{10} = V,$$

$$F - E + V = 2,$$

and

$$3V = 2E.$$

Hence $F = \frac{1}{4}V + \frac{1}{6}V + \frac{1}{10}V = \frac{31}{60}V$ and

$$2 = F - E + V = \frac{31}{60}V - \frac{3}{2}V + V = \frac{1}{60}V,$$

so $V = 120$ and $E = 180$, $F_4 = 30$, $F_6 = 20$, $F_{10} = 12$, the polyhedron being shown in Figure D.15.

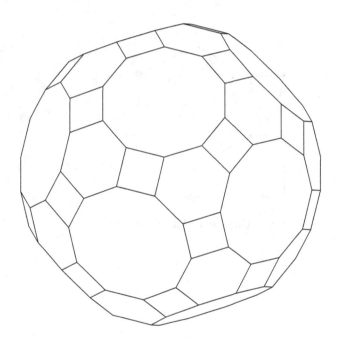

Figure D.15   Exercise 5.3

5.4 Now $F = F_4 + F_5$ and $2E = 4F_4 + 5F_5$, so

$$F_5 = 2E - 4F.$$

Also, $V = V_3 + V_4$ and $2E = 3V_3 + 4V_4$, so

$$V_3 = 4V - 2E.$$

The two derived equations give

$$8 = 4(F - E + V) = (4V - 2E) + (4F - 2E) = V_3 - F_5.$$

But the vertices of the pentagons occur one at each of the V_3 vertices with three edges, so $5F_5 = V_3$. Hence $4F_5 = 8$ so that $F_5 = 2$.

A pentagonal prism satisfies the conditions, as does any number of such prisms glued together by their pentagons.

5.5 Figure D.16 shows a hemisphere filled with four (topological) triangles, with two semicircular edges joining the north and south poles. A similar arrangement of triangles fills the opposite hemisphere.

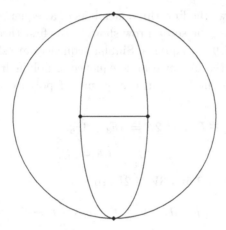

Figure D.16 Exercise 5.5

5.6 Our definition of polyhedron requires at least three faces at each vertex, so $F \geq 3$ and, as $3F \leq 2E$, we have $2E \geq 9$. Also $3V \leq 2E$ so, on the sphere,

$$F = 2 - V + E \geq 2 - \frac{2}{3}E + E = 2 + \frac{1}{3}E \geq 2 + \frac{3}{2}$$

and so $F \geq 4$.

5.7 (a) We have

$$F = F_6 + F_7 + F_8 + \dots,$$

$$2E = 6F_6 + 7F_7 + 8F_8 + \dots,$$

$$F - E + V = 0,$$

and

$$3V \leq 2E,$$

so
$$F = E - V \geq E - \frac{2}{3}E = \frac{1}{3}E$$
whence
$$6(F_6 + F_7 + F_8 + \dots) \geq 6F_6 + 7F_7 + 8F_8 + \dots,$$
and
$$F_7 + 2F_8 + 3F_9 + \dots \leq 0.$$
Consequently the polyhedron consists entirely of hexagons.

(b) In a polyhedron of eight or more faces, each pair being adjacent, every face would be seven-sided or larger. We have shown in (a) that this cannot happen.

5.8 Figure D.17 shows the first three members of a sequence of polyhedra with $4, 6, 8, \dots$ squares, the second row showing the first three members of a sequence with $5, 7, 9, \dots$ squares. Similar sequences are shown for hexagons. Also shown are the first two in a sequence of polyhedra with $8, 12, 16, \dots$ triangles and the first two in a sequence of polyhedra with $6, 10, 14, \dots$ triangles.

5.9 We have $F = F_3 + F_6$ and $2E = 3F_3 + 6F_6$, so
$$3F_3 = 6F - 2E.$$

Also, $F - E + V = 2$ and $3V \leq 2E$, so
$$2 = F - E + V \leq F - E + \frac{2}{3}E = F - \frac{1}{3}E$$
and
$$6F - 2E \geq 12.$$
Hence $F_3 \geq 4$.

Figure D.18 shows the truncated tetrahedron with four triangles and four hexagons. The hexagonal antiprism, with twelve triangles and two hexagons, is shown in Figure 5.4.

To construct polyhedra on the torus consisting of triangles and hexagons, start with the plane tessellation of triangles and hexagons, where each edge is an edge of a triangle and a hexagon. Figure D.19 shows two suitable polyhedra.

5.10 We have $F = F_3 + F_4$ and $2E = 3F_3 + 4F_4$, so
$$F_3 = 4F - 2E.$$

Also, $F - E + V = 2$ and $4V = 2E$, so
$$2 = F - E + V = F - \frac{1}{2}E$$

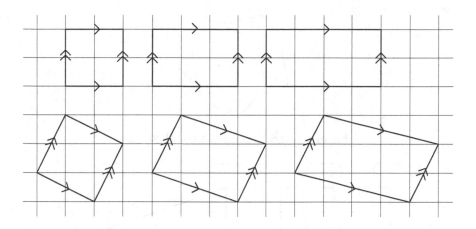

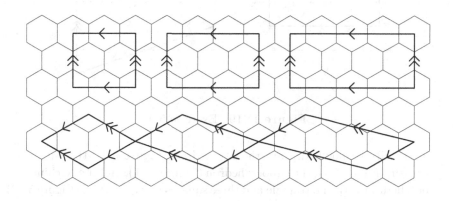

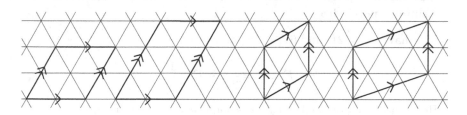

Figure D.17 Exercise 5.8

and

$$4F - 2E = 8.$$

Hence there are just eight triangles.

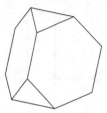

Figure D.18 Truncated tetrahedron

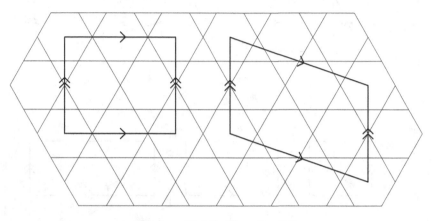

Figure D.19 Exercise 5.9

Four suitable polyhedra are the octahedron (Figure 5.4), the square antiprism (Figure 5.4), the cuboctahedron (Figure 5.21), and the Archimedean polyhedron with one triangle and three squares at each vertex (Figure 5.22).

For the polyhedron on the torus start with one of the plane tessellations that has two squares and three triangles at each vertex. Figure D.20 shows a suitable polyhedron.

5.11 We have $4F = 2E$ and $6V = 2E$, so

$$F - E + V = \frac{1}{2}E - E + \frac{1}{3}E < 0.$$

Hence no such polyhedron can exist on the sphere or the torus.

5.12 Suppose that we have a polyhedron on the torus with an edge between each of m vertices and each of the remaining n vertices. As each edge goes from one of the m vertices to one of the n vertices, every face must have an even number of edges, so $4F \leq 2E$. Also, $m + n = V$ and $mn = E$, so

$$0 = F - E + V =\leq \frac{1}{2}E - E + V = V - \frac{1}{2}E = m + n - \frac{1}{2}mn.$$

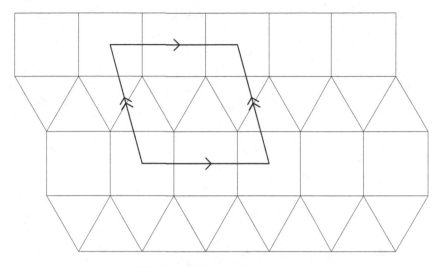

Figure D.20 Exercise 5.10

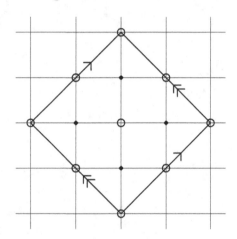

Figure D.21 Exercise 5.12

If $m = 3$, we require $n - \frac{3}{2}n + 3 \geq 0$, or $n \leq 6$. If $m = 4$, we require $n - 2n + 4 \geq 0$, or $n \leq 4$.

Figure D.21 shows a polyhedron on the torus with two sets of four vertices as required.

5.13 Each "side" of the polyhedron consists of three mutually perpendicular intersecting sets of square section tubes.

Bibliography

[1] Adams, C. C., *The Knot Book*, Freeman, 1994.

Ostensibly for the non-mathematician, this book brilliantly presents many of the outstanding open questions in the subject as an invitation to the reader to learn more.

[2] Armstrong, M. A., *Basic Topology*, Springer-Verlag, 1983.

This is an excellent main course in topology, which we particularly recommend to follow our aperitif.

[3] Bryant, V., *Yet Another Introduction to Analysis*, Cambridge University Press, 1990.

There are many books on this important but delicate subject: this is one of our favourites.

[4] Coxeter, H. S. M., *Introduction to Geometry* second edition, Wiley, New York, 1981.

Coxeter is a great geometer, and this is a classic.

[5] Cromwell, P. R., *Polyhedra*, Cambridge University Press, 1997.

The story of polyhedra is beautifully told here.

[6] James, I. M. (editor), *History of Topology*, Elsevier, Amsterdam, 1999.

This collection of articles mostly covers the development of the subject since Poincaré.

[7] Pont, J.-C., *La Topologie Algébrique des Origines à Poincaré*, Presses Universitaires de France, Paris, 1974.

Here is a very good book on the early history of the subject.

[8] Prasolov, V. V., *Intuitive Topology*, American Mathematical Society, 1995.

Prasolov is an inspired and inspiring writer.

Index

n-antiprism, 72
n-dipyramid, 72
n-pair, 22
n-point, 20–22
n-prism, 71
n-pyramid, 72

adjacent, 71
ambient isotopy, 113
Archimedean solids, 87

bijection, 33
bounded, 4, 48, 51, 57, 84, 108, 109

circle, 5, 12
circuit, 44
close subsets, 46
closed path, 93–97, 99
closed set, 7, 15, 41, 51, 57, 84, 108, 109
closeness graph, 45, 47
closure, 39, 41
compact, 85, 108
complement, 33
component, 18, 19
congruent, 1
connected graph, 44, 90
continuous, 1, 16, 46, 48, 93, 95, 98, 99,
    105, 107–110, 122
cross cap, 57, 84, 85
crossing number, 114
cube, 73, 74, 85
cuboctahedron, 146

cut-pair, 22
cut-point, 20–22

deformation, 26, 37, 97, 99, 113
disc, 5–7, 34, 35, 51, 53, 62, 70
dodecahedron, 72–74, 85
dual polyhedra, 85, 90

edge point, 7, 37, 51, 103, 104
enantiomorphic, 88
equivalence relation, 6, 18, 56
equivalent polyhedra, 73
equivalent subsets, 26, 33, 41, 47
Euclidean geometry, 58
Euclidean set, 1, 109
Euler characteristic, 83
Euler number, 83, 85
Euler's theorem, 74, 89, 121

fully symmetric, 86, 87

genus, 52, 84, 122, 124
graph, 44, 45, 47, 121

half-open, 41
handle, 52, 62, 84, 85
hexagonal antiprism, 144
hole, 52
homeomorphic, 1, 3, 6, 9, 15, 19, 21, 22,
    26, 33, 51, 54, 69, 85
homeomorphism, 1, 3, 26, 69, 104, 105,
    107, 123, 124
homotopy, 97

icosahedron, 72–74, 85
identification space, 56, 110
incident, 69, 71, 73, 75, 86
interior point, 103, 104
isomorphic graphs, 45, 47
isomorphic polyhedra, 73
isomorphic rooted trees, 47

joined, 44, 45
Jones polynomial, 113, 116, 117, 119,
    125

Kauffman polynomial, 117
Klein bottle, 53, 54, 58, 59, 62, 82, 125
Klein handle, 63
knot, 113, 125
– prime, 115
– trefoil, 31, 113, 116–118

Laurent polynomial, 116
link, 117

Möbius band, 38, 51, 52, 56, 93, 102,
    104, 124
magic, vii

neighbourhood, 39, 41, 51, 54, 56,
    105–108
non-orientable, 52, 54, 57, 58, 62
not-cut-pair, 22
not-cut-point, 20

octahedron, 59, 72–74, 85
one-sided, 52, 124
open, 41, 108, 109
orientable, 52, 62, 63, 85, 123
oriented knots, 116

path, 15, 16, 18, 43
path-connected, 15–17, 51, 85
Platonic solids, 73, 74, 87
polygon, 69, 70, 73, 74, 87
polyhedron, 69–71, 85
– regular, 71, 72, 74, 77, 87
– semi-regular, 87–89
– spherical, 71, 74, 89

real projective plane, 57–59, 62, 83
regularity, 85
Reidemeister moves, 114, 116, 117
ring number, 80, 81, 87
rooted isomorphism, 47, 48
rooted tree, 47–49

skein relation, 117, 118, 125
sphere, 8, 52
– n-dimensional, 9
standard polygon, 69
stereographic projection, 8, 11
– n-dimensional, 9
string, 44
surface, 51, 52, 57, 69, 71, 85
– with boundary, 51, 103
symmetry, 73, 88
– of Platonic solids, 87

tetrahedron, 72–74, 85
– truncated, 144
together, 18, 19
topological property, 3, 4, 15, 17, 108
topological space, 54, 56–58, 105, 108
torus, 11, 12, 35, 51, 52, 77, 110, 118
– (m, n)-circles, 103
– double, 52
tree, 43, 44, 46, 89, 90
two-sided, 52

unknot, 113, 115–117, 119